Qualification for Computer-Integrated Manufacturing

Springer
Berlin
Heidelberg
New York
Barcelona
Budapest
Hong Kong
London
Milan
Paris
Tokyo

Felix Rauner (Ed.)

Qualification for Computer-Integrated Manufacturing

With 48 figures

 Springer

Felix Rauner
Institut Technik und Bildung
Universität Bremen
FB 11/1TB
Grazer Straβe 2
28359 Bremen
Germany

ISBN-13: 978-3-540-19971-7   e-ISBN-13: 978-1-4471-3064-2

DOI: 10.1007/ 978-1-4471-3064-2

British Library Cataloguing in Publication Data
A catalogue record for this book is available from the British Library

Typesetting: Camera ready by editor

34/3830-543210 Printed on acid-free paper

Contents

Vocational Training for Computer-Integrated Production

Prof Dr. Felix Rauner
Institut Technik und Bildung, University Bremen
Bremen, Germany

Abstract

The purpose of the "CIM Training in Europe" Conference, in the eyes of its sponsors, is to present selected examples of vocational training for computer-integrated production and discuss them among experts. After a 5-year experimental phase the contours of various CIM training concepts are now beginning to take on definite shape. Four directions of development and fields of application can be clearly distinguished from one another.

1   The CIM learning factory

is an attempt to reproduce the reality of computer-integrated production at a very practice-oriented level. As a rule, the technical components of computer-integrated production, such as CNC machine tools, handling techniques, CAD equipment and simple PPC (production planning and control) systems, are integrated into the CIM learning factory. Other components, like transport facilities and warehousing, are implemented only in principle. The difference between a learning factory and a real factory is shown most clearly when one uses the real production process and the related work processes as a yardstick for comparison. In one case we are dealing with flexible production and in the other with "production of training". Only the work process itself can act as the teaching environment for the knowledge and skills required. However, few examples of this have emerged up to now. In many of the implemented learning factory examples those involved were content with singular linking and interconnection of CIM modules and the more or less automatic manufacture of simple parts or products. Two central questions are the subject of discussion here.

■   Are the problems and aspects of the introduction of CIM and permanent further development and modification of flexible production really reflected by the "rather inflexible and more singular" interlinking of CIM components for the manufacture of a simple part or product? Is it possible to reproduce the specific elements of flexible computer-integrated production in this way? The company-specific production technology, to be organised and implemented according to the respective special needs, may not be sufficiently taken into account in this process.

■   How great is the danger that simulation of a factory in a quasi-factory may suppress precisely those aspects that are important in differential implementation of increasingly open systems and system architectures of production technology:

technical flexibility, a variety of applications for company-specific implementation, openness and flexibility of work organisation as well as openness to industrial product and process innovations?

2 Models of computer-integrated production technology

reduce the complexity of computer-integrated production to the dimension of technology. As a rule, they allow for didactically structured, exemplary reproduction of complex technologies in easy-to-comprehend teaching systems and training aids. The focus here is again on singular interlinking of individual CIM components in the form of technical models. These CIM models, in addition to other training aids such as microelectronics and production-related components, become a separate product. There is usually no link-up with already existing CIM components, such as CNC-controlled machine tools, training robots, etc. In this area of CIM training the main questions relate to the realistic content of the technical models and the possibility of making essential aspects of the work process and work organisation part of the training. If it is right that the design of user interfaces between man and machine is of central importance for training requirements in flexible computer-integrated production and that there is considerable scope for shaping here, then it would be very important to provide for such scope for shaping and variety in the model technologically as well. In the tradition of technological didactics, however, the main emphasis is on the respective basic principle and how technology functions in principle. In this case a fundamental extension of the technical-didactic approach is required. Besides the principle, one must consider the dimension of variety of design as well as company- and work-related shaping of computer-aided tools and systems.

Simulation of computer-integrated production oriented to the technical aspect neglects the work-related dimension of CIM. There is no objection to that as long as it is possible to implement an integrated concept that incorporates the work-related aspects in a comprehensive CIM training approach. The extent to which this has been achieved in the projects presented within the framework of this conference will be examined in the respective part of the conference.

3 An open place of learning capable of migration

already suggests in its terms a concept that goes beyond the first two approaches. The objective connected with this approach is reflected, in particular, in three features.

■ The computer-aided components of flexible production or process automation existing in each case are opened to extensive, flexible interlinking and networking via correspondingly designed interface technology. In this manner the place of learning - by all means comparable to company organisation development - is to be set up as a process of learning site development. From an economic point of view, this would additionally save significant investment costs.

■ An open place of learning capable of integration is the most suitable approach for incorporating aspects of the work process and work organisation in the training concept in addition to the practice-oriented simulation of technical aspects of computer-integrated production.

■ The central feature of shapability of the individual components as well as of the technical interface and "work and technology" link-up can be implemented best in a learning site concept.

These three features have a definite affinity to the guiding principles of "participation" and "shapability", which are of central importance for modern vocational training. Regarding this approach, the question of whether it can be implemented with reasonable effort and expenditure in vocational training practice must be critically examined. Realisation of the interface quality necessary for this model requires a very high level of training with respect to technological science, work studies and vocational education on the part of developers and users. Furthermore, a prerequisite for broad application and situative use of a modular and flexible CIM workshop concept is that the time-related and organisational scope for utilisation of this variability is also available to the users.

4   Software simulation models for CIM

are, finally, another approach for implementation of CIM training. The consistent element in this approach is that the software side of CIM is emphasised in a particular way. The complexity of company activities can be represented elegantly through software technology. Other outstanding features are the pronounced virtual and experimental quality of many CIM software models as well as the comparatively elegant availability of this technology for users.
Therefore, it can be expected that the availability of CIM simulation software in supply-oriented training measures and courses will spread to a certain extent because of these features. In demand-oriented training concepts, on the other hand, it must be asked whether the simulation software not only has a representational function but also performs tool functions so that real company functions can be simulated and processed. In my estimation computer-aided work systems with a marked tutorial quality - systems which are work and teaching systems at the same time - will become considerably more important in the future. In this area of development it is crucial to ask where the limits of the simulatability of company production practice are drawn and whether or not there is a tendency to convey an implicit picture of full automation which clearly underrates the central question of training and the organisation of work and production. After all, such an approach might lead to a significant underrating of mechanical, hydraulic, pneumatic and electrical elements.
The examples presented in these four fields of application and development of CIM training during our conference offer us the opportunity of discussing the question implied here with the experts present on the basis of the varied experiences of the speakers and project participants. I hope that we are successful in conveying ideas and orientations for the further development of this promising area of vocational training through this assessment of the field of CIM training in Europe.
This is one of the great challenges for European vocational training research. Just

as the European ESPRIT program attempts to support Europe's competitiveness in the field of information technology beyond the limited possibilities of national research and development programs, it appears necessary to me to set up a European research and development program for vocational training in the area of flexible computer-aided production.

Right from the beginning of this European conference we are certainly already united by the common basic conviction that the shaping of the future European industrial landscape, which should be humane and competitive at the same time, will require a high level of general and vocational education. The potential of flexible computer-aided and integrated production technology is available to us worldwide, supported by globalisation in the development of basic technological innovations and the successful standardisation efforts of the ISO. The success in introducing flexible computer-integrated production which matches quality competition depends much more decisively on the factors of human resources and organisation of production. Education and training are thus becoming a key factor for the transformation of the production technology potential for competitive production adapted to regional and company-specific conditions.

A task-centered learning concept

Jürgen Broschat; Walter Kötter; Tillman Krogoll
CIM training factory, Berlin Adlershof
Berlin, Germany

The provides comprehensive training geared to market needs.

1   Current Situation

Production, cooperation, the labor market and personnel training should be given new form, since the structural changes in Europe have given rise to new economic, social and ecological opportunities and minimised associated risks /1/. This situation calls for broad-based concepts designed for long periods of time.

Industrial plants are either newly built or expanded all over the world. The highly developed industrial countries cannot remain competitive unless their standardised mass production of sophisticated products is highly automated. However, promising economic prospects are opening up in view of the vast demand for high-grade technical products. The control mechanism governing the markets for such products primarily works via product quality and degree of innovation rather than price. The new requirements arising in terms of flexibility, quality, productivity and innovation can be met most successfully using those production systems that are based on computer-aided manufacturing technology associated with highly qualified work /2//3//4/. The main features of such systems include the following:

-       renunciation of complete automation in favour of human decision-making with a low degree of formalisation of decision-making and control structures;

-       decentralised work organisation involving flat hierarchies and a strong trend towards delegation of planning, supervisory and control functions to subordinate levels, especially the shop floor;

-       highly reduced division of labor with workers and equipment pooled around the task to be accomplished /5/.

The more such flexible work systems develop and spread, the greater the tendency for a bottleneck to form with regard to human resources. If qualified workers are to be available in adequate numbers, redundant workers should no longer be left unemployed but should become involved in advanced training schemes. Unemployment today will cause a lack of qualified workers tomorrow.

As CIM is being discussed as one of the major technical and organisational advances of recent times, many companies are redoubling their efforts toward raising the skill level, especially of those workers who are still lacking an appropriate level of training.

In the eyes of teachers and psychologists alike, there are few good concepts oriented to basic and advanced training in CIM environments. A BMFT-AuT project entitled Development of an Extended Task-Centered Learning Concept for Experience-Based Activity as Part of Computer-Aided Work /6/ aims at creating learning processes which are based on the experience accumulated by trainees, help them grow into the new environment and extend their wealth of experience through self-reliance. Although the project can be considered a good basis, it is restricted to the manufacturing process itself. It relates only indirectly to the learning processes that are also required in the technical office and assembly shop. The same applies to the necessary process of teaching trainees to participate and cooperate beyond department boundaries, without which company-wide task integration would be impossible.

2   The Challenge: A Task-Centered Training Factory

The factory as the place of instruction has a number of advantages over conventional courses which are offered by manufacturers or in general advanced training schemes along subject-related rather than learning-related lines. These advantages arise out of the fact that the trainee is directly associated with current and future tasks at the workplace and can draw on his background as a production worker. This should be complemented with an appropriately organised learning process and with didactic and methodological means geared to the needs of adults.

However, there are quite a number of reasons why the positive results obtained from task-centered learning should not be restricted to factories but be used in general training schemes as well.

Setting up a properly structured task-centered training factory and conducting accompanying research and development for the introduction of integrated task-centered learning systems and appropriate organisational and technological solutions is an important step designed to bring about innovation in this field.

Major aspects in the process include the following:

■   A training factory based on task-centered learning systems suits the purposes pursued much better than does a real factory. All major elements required for learning at a factory can be provided there. In addition, it offers advantages that make the relatively high expenditure justifiable:

-    The production run is only a secondary consideration. In many cases, it is more economical and makes it easier to provide the necessary setting for decentralised learning.

-    When it comes to familiarising trainees with fundamental aspects, specific problems can be presented very systematically.

-     Generalising and transfer effects are easier to achieve in this way.

-     Considering production efforts only as the secondary factor, trainees will learn much more effectively, without being disturbed or interrupted.

-     Newly arising problems can be taken up as the need arises.

■   A training factory is very important or even indispensable, if training and task-centered learning systems are to be developed, tested and transferred. Technological and organisational alternatives for workshops and production buildings can be tested, optimised and properly converted into task systems. At the same time, they are ideal as references or for consultancy, especially for small and medium-size companies.

■   Countries with recent economic changes, such as those in the eastern part of Germany and southeastern and eastern Europe, are witnessing a gigantic demand for high-quality training courses geared toward the needs of the market now and in the future. People there are called upon to adjust their behaviour to the conditions of free enterprise and adapt to a work environment which has changed in both organisational and technical terms. Companies where job-oriented training should be provided do not exist or are themselves dependent on comprehensive assistance and support. Training factories, with their future-oriented training programs, have an important impact on the labor market and may give regional economic recovery a major boost.

■   These factories are the only alternative when production is to be launched at new sites. From the viewpoint of the organisation that provides advanced training, learning in such a factory could be a decisive preliminary step toward continuing on-the-job learning organised decentrally at the company level.

Preparations for setting up a CIM-based training factory are under way in Berlin's Adlershof borough. The project holds the unique promise of becoming a model for supracompany advanced training whose attraction will reach beyond eastern Germany well into its western part and the rest of Europe.

The CIM training factory in Berlin Adlershof uses the technical, organisational and human resources of the Scientific Instrument Manufacturing Centre which used to be part of the former Academy of Sciences of the GDR and which is being liquidated at present. It is being set up at the initiative of the Centre's CAD/CAM Department. Since 1990, courses lasting several months and held under the Labor Promotion Act have been provided at the company's CAD/CAM training centres, with participants being mainly

- technical and commercial staff,
- technicians, engineers, business economists,
- graduates from vocational schools and universities.

The CIM training factory will gradually be set up and the task-centered learning systems developed from 1991 to 1994. The aim is to qualify people for self-reliant, integrated work and the safe use of conventional and computer-aided equipment in single and small-batch production processes in the areas of engineering, instrument and plant manufacturing.

The course is based on several months' instruction involving

- work-oriented advanced training provided in production and work islands.

In addition, it will be necessary to introduce advanced training courses of shorter duration for all the different sectors within the factory:

- Instructor training for those with an industrial background to develop task-centered teaching and learning concepts suitable for technological and organisational innovation.

- Workshops, introductory seminars and courses lasting up to four weeks and dealing with the humane design of work in CIM environments.

- Company- and problem-related one-day or one-week seminars providing training to staff of the lower company echelons or company or union representatives.

3 Task-Centered Learning System

The aim is to develop, test and evaluate a generally applicable advanced training pattern covering everything from the technical and commercial office to production and assembly. It is designed to be a system comprising various subsystems of tasks to be tackled by the trainees within human-centered CIM structures /7/.

The envisaged factory and the accompanying research project will involve a highly cooperative task structure consisting of a design island, a planning and logistics island, and three production and assembly islands (Fig. 1). At the islands, trainees will learn in appropriately developed task-centered learning systems how to approach ambitious single tasks in a work environment representing a real-life mixture of conventional equipment and suitable CIM components. This is to be followed by practicing the process of self-regulation in a working group using a task-centered learning system for the central tasks /8/ common to a particular island. In addition, the central task will involve cooperation between the design, preproduction, production and assembly stages. The entire learning and work process will be planned, structured and backed up by a "permanent staff" who will have to act with a high degree of personal and job-related expertise. Their career background may range from research worker to development engineer. They will be taught teaching and didactic skills and be incorporated into the accompanying research project for which funds still have to be applied. The main aim of the research project is to design the tasks, especially during the early course phases, at a level where neither too much nor too little is asked of participants. In general, the skeleton conditions for practicing the training tasks must neither be regulation hindrances nor excessive regulation requirements /9/.

Another prominent feature of the factory is its coherent simulation of real-life conditions in that customers' orders are just as involved in the learning process as are the preparation of orders and development and design tasks. It even includes the production and assembly of the developed product.

This "Learning under Market Conditions" even stretches beyond the proximity to real life achievable with the task-centered learning concept because it lowers the trainees' threshold of applying what they have learned to their future jobs - a weak point of all supracompany advanced training schemes. As far as it is possible, considering the current situation in the labor market, the process of transferring accumulated knowledge to new jobs becomes part of the evaluation carried out under the research project. If possible, this process would have to be supported by follow-up training at the workplace. However, auspicious transfer conditions can be expected not only because of the activity-oriented logic learning and the comprehensive simulation of practice underlying these courses, but also because of a number of other aspects:

■ The complicated nature of the products made by the company (ZWG) in the past, the expertise of the teachers to be employed and the special demands made on product quality and customer satisfaction in the realm of scientific instrument manufacturing will make it possible to confront trainees with challenging problems and include the need for innovation, particularly on the part of companies in East Germany, in the teaching programme.

■ As the level of difficulty progresses with the various task-centered learning systems, trainees go through a wide variety of qualitatively differing requirement constellations. The scheme involves a pattern where trainees go through a sequence of qualified self-reliant work, interdependent cooperation and partly autonomous, self-regulating group work. These building blocks are part of the system of task-centered learning subsystems to be developed and enable trainees to transfer what they have learned to a wide range of organisational environments. Furthermore, course participants will be able to assess these environments on the basis of their own needs and the criteria of human-centered work design. Last but not least, they will be able to participate in innovation processes at their company.

■ As trainees become more and more familiar with various types of tasks and equipment, they will be able to consolidate their knowledge and skills and arrive at more generalised conclusions. In addition, they enhance their scope of experience concerning in-house interrelationships and extend their ability to compare work tasks, work structures and equipment configurations. Since differing types of equipment are used within one and the same department of the factory - which is rather expensive and indicates a tendency toward complicated data integration - this can be further enhanced, though only at a high level of task-centered learning systems.

As far as the methodology of the project is concerned, it has been developed according to a prospective job design scheme /10//11/ based on techniques called VERA, VERA-G and RHIA along with the KABA Guidelines, which are nearing completion, taken into consideration. The entire structure of the CIM training factory is to feature an ideal operating and organisational setup in accordance with "human criteria" of work design. With such an ideal task structure in the factory, the teaching staff, while undergoing further education and working under the guidance and supervision of project managers, will have to develop an advanced training scheme encompassing the entire factory as a system of task-centered learning subsystems. The underlying didactic concept will be based on the CLAUS method /7/. The following aspects are to be given particular attention during the process of task-centered learning system development:

■   Consistent use should be made of the mixed availability of conventional and computer-aided equipment at the place of instruction. Participants should be able to assess alternative technologies and their impact on job design and workers' scope of action.

■   Ample scope should be given to the search for suitable ways of referring to trainees' experience with conventional equipment and to special training forms for the transfer of such knowledge to computer-aided work - not only in the workshop but in technical and commercial departments as well.

■   Consideration of various learning strategies, taking into account differences between young and old trainees.

4   Equipment of the CIM Training Factory

Specifications for selecting and designing appropriate equipment are drawn up on an interdisciplinary basis under the leadership of work designers. These specifications should be specific, concrete in terms of system assessment and development, and practice-oriented.

The CIM components to be installed will be chosen so as to induce favourable conditions for both the learning process and transfer to future work tasks. The following criteria will be of primary concern when selecting the CIM components to be used as teaching aids and operating equipment:

■   All components must be suitable for the tasks to be performed by the trainees while learning and working.

■   Future-oriented working practices necessitate the availability of equipment and installations which encourage group work and provide scope for experience-oriented action.

■   Compatibility between the components to be installed or, instead of that, utilisation of a component as an appropriate means suited for use with several tasks.

■   The component should be promising, though "maturity", which experience has shown to be highly important in terms of task appropriateness, must not be neglected.

-        The degree to which such components are prevalent - to ease transfer.

-        Flexibility regarding varying requirements, i.e. these components should lend themselves to as many applications as possible.

The CIM training factory provides a realistic field where currently available CIM components can systematically be selected and optimised and different components used for one and the same range of tasks to be compared. In addition, it may initiate the development of new products for the two integration paths of CNC/CAD and WSL/PPS, have prototypes of such products installed at the factory and have them gradually developed into systems that can be brought on the market (evolutionary system development). In both production and assembly islands, NC programming as well as the planning tasks at the control panel will be based on the use of a standardised user shell and involve man-machine dialogue based entirely on work psychology criteria. Trainees are to be taught the way in which the system interfaces of horizontal integration, for example, between design and manufacturing, are realised, in both directions. But the emphasis is on the link between CAM and CAD. This will be reversed insofar as the full availability of CAD and PPS data on the standardised workshop terminal takes precedence over other criteria of interface design. If possible - to encourage inter-departmental cooperation by way of "computer-supported cooperative work" (CSCW) - a uniform product data model is to be developed which, again entirely aligned to work psychology criteria, should contain all geometric, technological, material-related and economic data necessary to carry out a particular task, without eliminating the possibility of structuring the system and without depriving it of its suitability for the individual task concerned.

CIM component manufacturers are expected to cooperate in terms of "evolutionary system development". Action-oriented bench-mark tests /12/ will be used to test whether or not a component is suited for a particular task. In addition, such tests will be made applicable to all ranges of tasks in the CIM training factory and be offered as a service to CIM user companies.

5   Prospects

Being work-related and human-centered, the methodology and the target group of the CIM training factory differ clearly from the 21 CIM transfer centres in the western part of Germany. However, the project will ideally strengthen the work-centered component of CIM technology transfer to be built by an association called MUT in East Germany through a joint application model.
The project of the CIM training factory in the Adlershof borough of Berlin has met with wide and positive response among research centres, companies and the manufacturers of production and information processing hardware. This was demonstrated not only by two workshops held in Berlin in mid-July 1991 but also by the numerous talks conducted since then. Representatives of renowned research and educational facilities will participate in the project. Researchers, business leaders and educationalists will be forming project-related associations this year.

Implementing this CIM training factory is an extraordinary effort to enhance the attractiveness of Adlershof as a hub of scientific and business activity in the German capital Berlin.
We wish to establish cooperative relations and an exchange of views with similar projects in other regions in Europe.

References:

/1/   Lehner, Fr.: Strukturwandel: Neue Formen der Produktion, der Kooperation, des Arbeitsmarktes und der Qualifizierung. Institut Arbeit und Technik 1989/90; Louisgang, Gelsenkirchen, Mai 1991.

/2/   Bullinger, H.-J., Fähnrich, K.-P.: Informationstechnik in der Produktion - Schwerpunkte zukünftiger Entwicklungen. 14. Europäische Congressmesse für Technische Kommunikation (14th European Congress Fair for Technical Communications); Plenarvortrag; Hamburg, 04/05 - 08 February 1991; ONLINE GmbH, Velbert 1991.

/3/   Martin, T.; Ulich, E.; Warnecke, H.J.: Angemessene Automation für flexible Fertigung. wt-Werkstattstechnik 78, 1987, pp. 17-23 and 119-122.

/4/   Brödner, P.: Fabrik 2000. Alternative Entwicklungspfade in die Zukunft der Fabrik. Edit. Sigma Bohn, Berlin 1985.

/5/   Gottschalch, H.; Hämmerle, E.: Arbeit-/sozialwissenschaftliche Erfahrung zur Gestaltung einer Werkstattsteuerung für Fertigungsinseln in CIM-Strukturen - Ergebnisse eines ESPRITS-Projektes der Europäischen Gemeinschaft. AWF-Fachtagung Fertigungsinseln, 6 - 7 Dec 1989; Bad Soden/Ts, Tagungsband (Minutes).

/6/   BMFT-AuT-Projekt: Entwicklung eines erweiterten Lernaufgabenkonzeptes für erfahrungsgeleitete Tätigkeitsanteile in der computergestützten Arbeit. FKZ. 01 HG 169.

/7/   Krogoll, T.; Pohl, W.; Wanner, C.: CNC-Grundlagenausbildung mit dem Konzept CLAUS. Didaktik und Methoden. Schriftenreihe "Humanisierung des Arbeitslebens"; Band 94, Campus Verlag, Frankfurt am Main 1988.

/8/   Gohde, H.-E.; Kötter, W.: Über die Bedeutung der Qualifizierung für die Kernaufgabe bei der Herausbildung von Arbeitsgruppen in Fertigungsinseln. Institut für Humanwissenschaft in Arbeit und Ausbildung der Technischen Universität Berlin IfHA-Berichte Nr. 22, Berlin 1989 (Fotodruck).

/9/   Leitner, K.; Volpert, W.; Greiner, B.; Weber, W.G.; Hennes, K.: Analyse psychischer Belastung in der Arbeit. Das RHIA-Verfahren. Handbuch und Manual. Köln, TÜV Rheinland, 1987.

/10/   Ulich, E.: Arbeitspsychologie Zürich: Verlag der Fachvereine; Stuttgart: C. E. Poeschel Verlag, 1991.

/11/   Kötter, W.; Volpert, W. u.a.: Prospektive Arbeitsgestaltung in der flexibel automatisierten Fertigung. Arbeitswissenschaften 34 (1990) 4, Verlag Die Wirtschaft GmbH Berlin.

/12/   Kötter, W.: Arbeitsaufgaben und Bewertungskriterien bei CNC-Fräsmaschinensteuerungen unter besonderer Berücksichtigung der Eignung für die Werkstattprogrammierung. Studienarbeit, Technische Universität Berlin, 1988.

Criteria of basic training in the field of "Computer Integrated Manufacturing (CIM)"

Dr. Michael Burgmer
Department of Mechanical Engineering
and Technical Didactic, University Dortmund
Dortmund Germany

1 A new keynote of industrial development: Computer Integrated Manufacturing

1.1 Economic and technological view

Obviously the structure and course of production processes are governed by the requirements of economic and technological efficiency. These take precedence over other necessities and also claim to make optimum use of the employed capital.

The current situation of European industry is primarily characterised by high personnel costs, low personnel flexibility, imposed environmental requirements as well as small national markets with varied norms and regulations. At the same time industry depends on export, which often has more than a 50% share of turnover. On home and international markets industry faces worldwide competition, thus forcing it to improve the effectiveness and reactions of business as a whole.

CIM is supposed to play an eminent role in assuming this long-term responsibility. CIM (Computer Integrated Manufacturing) stands for a broad range of industrial development aimed at the integration of three flows: information, material and resources.

CIM is a concept for production in which products are specified according to market needs, designed by means of computer support, converted into digital production data and then distributed.

The technical framework applies to concepts of process management (e.g. principles of flow, continuity, interlacing) and process intensity (e.g. increase of active energy, minimisation of expenditure). The economic framework applies to concepts like "just-in-time" and "zero inventory".

The performance of the production systems is affected by the progress of communication and information technologies.

To summarise the problems mentioned in Fig. 1, producers of all company sizes seek to

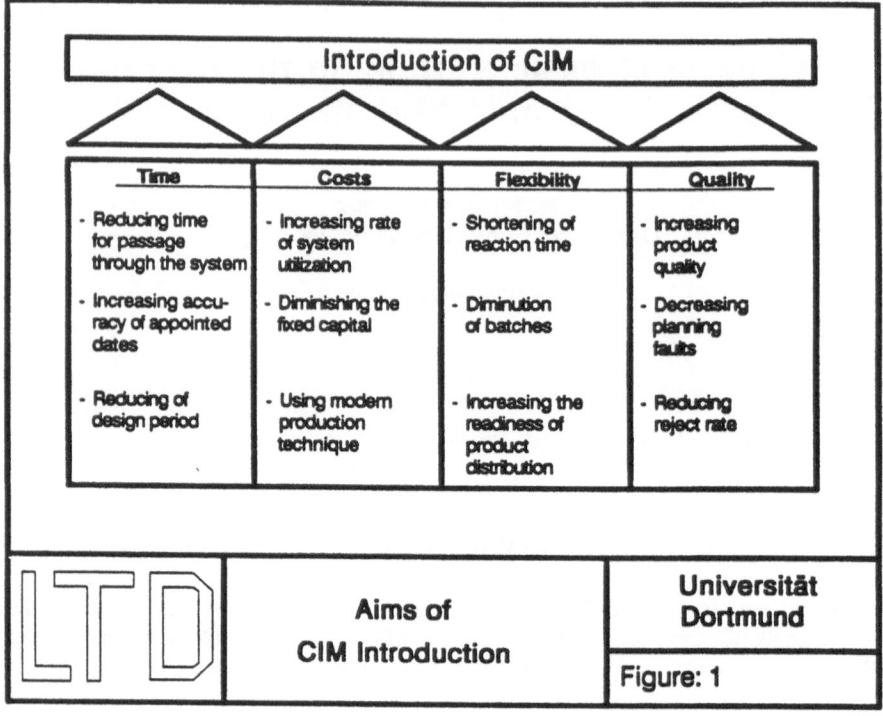

Time	Costs	Flexibility	Quality
- Reducing time for passage through the system	- Increasing rate of system utilization	- Shortening of reaction time	- Increasing product quality
- Increasing accuracy of appointed dates	- Diminishing the fixed capital	- Diminution of batches	- Decreasing planning faults
- Reducing of design period	- Using modern production technique	- Increasing the readiness of product distribution	- Reducing reject rate

Introduction of CIM

LTD · Aims of CIM Introduction · Universität Dortmund · Figure: 1

Fig. 1 Targets by planning CIM

- reduce the time for developing, manufacturing and delivering the products,
- decrease the costs by increasing the rate of system utilisation, by diminishing the fixed capital and by using modern production techniques,
- increase flexibility by shortening the reaction time, by reducing batches and increasing readiness for distribution of the products,
- increase the quality of the product, decreasing planning faults and reducing the reject rate.

To achieve these improvements, companies introduce computer controlled integrated systems.

1.2 A View of Anthropomorphism

The hitherto existing view only is partial. All human actions within the scope of economic processes are carried out according to certain laws of nature and within a defined time frame. Robert Mayer's definition of the first law of thermodynamics made this point clear: the direction taken by natural occurrences is not reversible. All high-quality and usable resources are finite.

Though the total amount of energy and matter remains constant, the quantity of usable matter and energy diminishes in each process of conversion of matter and energy.

Human behaviour - like nature - inevitably leads to elimination of order and increase of disorder.

Accordingly, all human actions within the framework of economic processes represent nothing more than the expansion and acceleration of this transformation. So the idea that man can create a "world of order" with the help of economics and technology is erroneous /1/.

Consequently, the claim that man has the competence to shape the world is relative. There is no denying that this transformation process has to be extended through far-sightedness, intelligence and benevolence and that it is relevant to improve the efficiency of this process. There are indicators that show that Computer Integrated Manufacturing can live up to this claim. If this were so, CIM training would gain additional fundamental importance.

1.3   Utilisation   of   computer-integrated   production   and development trends

Integrated systems of production technology are of considerable importance for economics in two respects:

-        for the producers of capital goods as a future product or a product component
-        for the users of integrated systems as a means of production in order to achieve competitive advantages.

The need for integrated production systems will increase in all different areas of business. CIM has a good chance of becoming an independent product in the next few years.

There is one risk involved in this development, i.e. the demand for stand-alone machines might decrease. At the moment and in the near future integrated production systems will only be available as customer-oriented solutions.

Three typical features characterise the capital goods sector

-        the above mentioned customer-oriented solutions
-        technical support for these customers with high quality standards while the machines are working
-        high-quality products.

Great profit can be gained from the just mentioned qualities for CIM as a European export product.

There are also shortcomings, however. The standard components required in integrated systems are weak points in European machine and plant construction. On the market there is a demand for development and implementation of individual solutions, but not for the production of standard components. Japanese and American manufacturers have, in some cases, achieved a very high production rate with standard solutions at reasonable costs. Thus in the near future we will have to take into account disadvantages in costs and competition for European products as well as a strong influence on norms and interfaces.

When looking at the risks on the way to obtaining CIM as an export product, we have to consider the additional production capacities built up worldwide in recent years. In Southeast Asia, for example, enormous capacities for production of machine tools have been established. This measure has led to a permanent change in the market situation. The reason for further risks is that the product value of integrated systems has increased considerably compared to business with stand-alone machines.

Japanese companies offer mass-produced products of an acceptable quality, but neglect the individual wishes of the customers. In comparison with this Japanese market strategy European businesses have to aim at markets with a smaller number of pieces, but great demands on individuality and technical quality.

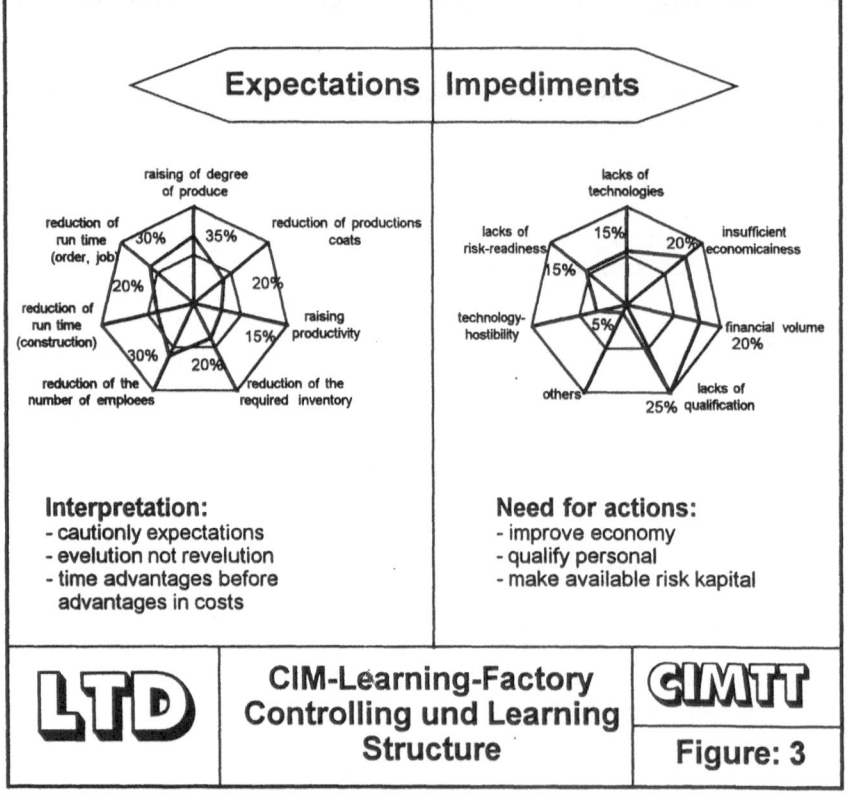

Fig. 2   Conclusions from the Delphi survey

In the near future more customised products will be in demand, both for capital goods and consumer goods. Shorter product and planning cycles together with more individual products will require a considerable time reduction in the fields of planning, product development and completion of orders. Integrated systems do meet these requirements because of their production flexibility and quality advantages based on automation.

In summary, we can say that expectations and obstacles referring to Integrated Production Systems can be quantified with the help of relevant investigations (see Fig. 2).

Surveys assessing the potential of CIM for the next ten years are very cautious, indicating that CIM will bring about an evolution rather than a revolution. Time advantages are expected, not only in production but also in development and design. As a consequence appropriate reductions of the required inventory will become necessary. According to experience up to now, the predicted advantages in costs are less important.

Four impediments to progress in production technology are cited. The most relevant one is the lack of personnel training (25%), followed by insufficient economy and technical shortcomings of technologies and tight financial resources. Remarkably small obstacles are seen in the rejection of technology and insufficient state support /2/.

The need for action shown in Fig. 2 can be deduced from the survey results just mentioned.

In the following further discussion of personnel training will be presented.

2 Criteria for a concept regarding training and further training in the field of "Computer Integrated Manufacturing"

2.1 Training demands

A general problem involved in automation of manufacturing processes is the connection between components that work differently, from the discharge of the unmachined material up to product packing, on the one hand, and ensuring a continuous material flow, on the other.

Let's take a closer look and include the following elements in the automation chain: product planning, purchase of the material, the design on the input side as well as the distribution of goods on the output side. This would then represent a CIM factory with perfectly linked computer-aided execution systems.
The dissemination of the CIM technology has only just begun.
However, the rate at which conventional factories incorporate CIM process will not only depend on machine- and information-related innovations but also on the speed of human mastery of the systems.

Up to now employees have not been trained in this field. Therefore, it is of eminent importance to promote fast, efficient training of personnel at all human working levels with regard to decision-making, planning, design, and operation. This means that constant further training and implementation of technical innovations have to be part of a long-term operational strategy.

The introduction of CIM technologies involves a change in training and an increase in training demands. This can be concluded from the following requirements for work using flexible manufacturing systems /3/:

-      the variety of information to be processed will increase;

-      the individual must be able to orient himself in a cross-linked complex system with organisationally fixed divisions of labour and team-work structures;

-      a cross-linked manufacturing system demands high standards concerning the understanding of technical connections and the integration of system and process knowledge;

-      the high manufacturing performance of these systems inevitably increases the decision-making scope and responsibility of the individual;

-      the cross-linking of different sub-systems increases the complexity of the problems to be solved;

-      every employee having to do with computer-integrated manufacturing systems must show a great deal of understanding for the repercussion of his action or non-action on the whole factory /4/.

2.2   Objective "ability to act"

The change in qualifications is characterised by the fact that there is a shift in the necessary training efforts for additional or different types of qualifications.
Subject-specific qualifications are normally tied to the corresponding products and the place of work and are naturally subject to technical and economic change. The innovation and extension of those qualifications will continue to be necessary after relatively short periods of time.
Vocation-specific qualifications of many occupational groups are so fundamental in some cases that they can be regarded as a basis for various kinds of specialisation. Such qualifications are often much less influenced by technical progress and will thus not lose their applicability very rapidly.

Extra-functional qualifications or key qualifications are of quite a general nature: on the one hand their value can hardly be diminished by technical development and, on the other, they are required to handle necessary adaptation processes.
In this context key qualifications are defined as the ability to combine sensory, cognitive and motivational components.

The sensory component includes the fully developed ability to absorb and assimilate information and, most importantly, to perceive and classify visual objects; to be more precise: to allocate these objects to the stock on hand in the sensory memory device.

When defining key qualifications, the cognitive component, i.e. the availability of a broad, well-developed inventory of heurisms, is of great importance. First of all, this means the complete availability of elementary cognitive processes:

- differentiating and generalising or comparing, which means comprehending differences and common features between comparable objects;

- ordering, which means producing objects with regard to one or more characteristics;

- abstracting, which means understanding essential and ignoring nonessential characteristics;

- generalising, which means comprehending common and essential characteristics when looking at several facts;

- classifying, which means assigning an object to a class or rank;

- putting in concrete terms, which means transferring and applying a universal principle to a particular problem.

Moreover, it is necessary to be able to combine these elementary cognitive processes to methods of solution (heurisms) for particular demands (problem, task/job) rapidly, accurately, independently and actively. Particular links to separate cognitive elementary processes, for example thinking in analogies and models, seem to be relevant for the solving of implosion problems; this is called "comprehensive thinking".

When defining key qualifications, the importance of the motivational component should not be underestimated. Motivation is the momentary willingness of an individual to direct and coordinate his sensory, cognitive and motor functions towards the achievement of an aim.

A motive is a certain orientation of the possible mode of action. With reference to key qualifications, particular motives seem to be linked with particular scales of values.

All training and further training programmes for Computer Integrated Manufacturing must adjust to this objective "ability to act".

2.3 Occupation-specific basic training in the CIM Learning Factory

2.3.1 Basic didactic demands

With regard to personnel conditions, impediments and expectations are linked to basic demands for safety devices of a high quality in a concept of training and further training in the field of CIM. The following can be pointed out:

- Maintaining consistency of subjects and sequences by harmonising the substance and extent of the teaching modules.

- Maintaining teaching effectiveness through didactically reduced information on the subjects in the learner manual, initiation of action and problem-oriented exercises and didactically commented instructions on every module for the instructors.

- Maintaining user-friendliness by considering psycho-physiological parameters of learning processes, e. g. for the layout of computer user interfaces.

- Inducing motivation by applying several coordinated teaching and learning methods, e. g. by dealing with texts individually through conventional lessons, interactive learning as computer based training, performance tests of the models of the CIM Training Factory and simulation of functionality faults.

- The modular performance of the CIM Training Factory allows one to instruct several groups of learners with different initial knowledges and learning abilities. The total curriculum includes more than fifty percent practice. This means that the learners are working on particular problems of the CIM Training Factory most of the time.

- The concept has to be effective and economical, i.e. the aims must be achieved at acceptable costs within a relatively short time.

- The concept has to be flexible with regard to technological, organisational, personnel and didactic changes.

- The concept includes the invariant, indispensable, fundamental and exemplary contents which are most important today and in the near future.

2.3.2 Basic structure of the CIM Learning Factory

Four years ago the Department Technology and Education of Dortmund University, subsidised by the European Economic Community in the COMETT programme, started developing and testing a multi-media system ("The CIM Training Factory") for basic training in the field of computer-integrated manufacturing. This system was intended to show the essential functional fields of an integrated system, including its cross-linking at a functional model level. The development and testing of teachware that meets the different levels of requirement of the target group of learners was associated with the development of the functional models

The CIM Training Factory consists of

(1) a well operating "model factory", where activities like job management, production control, design, manufacturing, including loading, material transport and assembly, as well as quality control and warehousing are flexibly shown in functional models and are controlled by means of cross-linked computers and

memory-programmed controls (MPC); during training the cross-linked computer structure is used like a language laboratory, see Fig. 3;

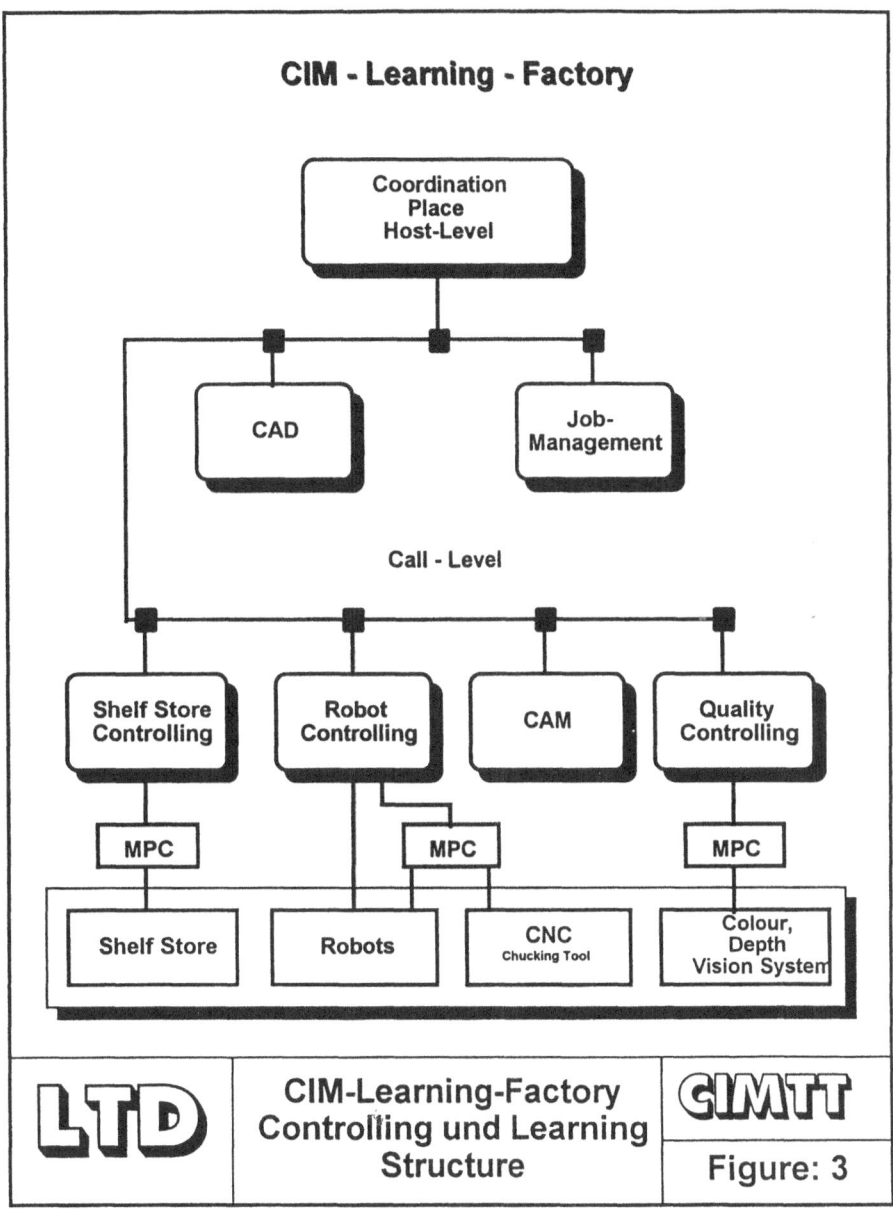

Fig. 3   CIM Lerning Factory - Control and teaching -

(2)   two different "teachware packages", the first for the target group of designers and decision-makers, the second for skilled workers and plant management. Both packages contain a reader and practical instructions for the trainees and an instruction manual for teachers and instructors. The latter book includes all the training media (transparencies, slides, video sequences).

Such a modular, flexible CIM Training Factory should succeed in carrying out the outlined demands, but it requires basic CIM training. It includes the particular advantage that the training aims are achieved by means of up-to-date teaching methods in an action-oriented and highly motivating way.
Training for blue-collar workers or plant management might include, for example, the detecting of simulated faults in the plant; for designers and decision-makers training dealing with variations of work organisations would be useful.

The training factory consists of at least three levels a workshop or system level, a cell or master computer level, and a coordination level.

At the head of the CIM Training Factory and within the scope of action of each trainee there is the workshop or rather the system level consisting of the following components:

(1)   CAD/CAM workstation consisting of a CAD workstation and a small industrial X/Y/Z CNC milling machine with pneumatic chucking tool for workpiece support;

(2)   Robot workstation consisting of a small industrial robot with 5 degrees of freedom (without hand);

(3)   Material flow system and quality control system consisting of an endless apron conveyor, driven with electrical motor, shunts, lifting and indexing units, pick-and-place robots for material change and quality control systems like colour measuring system, depth measure system (tracing pin) and vision system;

(4)   Shelf store model, fully operative, with 32 cases, shelf conveyor, sensors and actors.

The cell level in the CIM Learning Factory consists of computers and the connected memory-programmed controls (MPC) for the station material flow, quality control, manipulating modules and shelf store. The coordination level includes computers used for the industrial planning and control system and job management as well as the host computer. This computer, also called "master position", allows you "to spy" into every cell computer. The computers at the cell or master level are coupled with the computer at the coordination level.
An essential advantage of the CIM training factory is the clear realistic representation of the material flow and processing stations as well as the fact that technological functions and business administration functions are combined.

2.3.3 Target groups

Skilled workers, master craftsmen/foremen and engineers in the fields of metalwork

and electrical engineering are involved in Computer Integrated Manufacturing. The requirements include:
basic knowledge of PCs, MPC, hydraulics and pneumatics.

A maximum of 12 trainees per course is recommended.

2.3.4 Aims

In detail the CIM Training Factory should contribute to the following aims:

(1)   enable consistent transfer from the model level to reality;

(2)   orient itself to industrial standards;

(3)   enable training on subsystems as well as on the overall CIM process;

(4)   make it possible for the trainee to get to know and organise his scope of planning and responsibility at his place of work;

(5)   take into account the individual receptiveness as well as the psychological and physiological limits of the target group;

(6)   enable various social forms of basic and further training, for example, individual learning with the help of computer-supported training sequences in solo learning boxes and as teamwork with a partner or in small or large groups;

(7)   offer various methods of basic and further training at every technical employee level, for example, conventional methods as well as role examples and simulations;

(8)   orient itself to future training profiles and cover all levels of training;

(9)   initiate extra-functional areas of competence and key qualifications.

2.3.5 Learning contents and its conception

CIM technologies are being introduced to an increasing degree. This means that training for the described types of qualifications must take place by means of highly integrated training systems. Extremely complex plants are almost exclusively used for the production process and are thus not available for basic training, just because of the expenses.

The same arguments that prohibit the basic training of pilots in actual practice are relevant for many production plants: one cannot afford a "crash".

Therefore, it is sensible to simulate automated production processes.

The CIM training factory combines basic and further training with the simulation of production processes along with an excellent demonstration of the material flow in

an extremely flexible manner. At least two ways of using the CIM training factory suggest themselves. At the first level the cell computers are not coupled with each other. This level allows you to become acquainted with some components of computer-aided manufacturing, for example, robots, NC-controlled axles, sensors, simulation, CAD with a large amount of practice, and computer-supported training. At the second level the cell computers are coupled with each other via CIM. By running the plant, the trainee will be able to improve his systematic thinking and his understanding of comprehensive connections. The training might include the detection of faults as well as the determination of strategies for eliminating those faults. Error and wear simulation is also possible. Within the scope of basic CIM training decision-makers, general CIM beginners, and plant managers are possible target groups for training in the CIM training factory. The latter particularly takes into account the situation in practice, for example, with the help of interactive computer-based training.
The following slides demonstrate this.

2.3.6 Configuration

(1) Media systems

Regarding the prerequisites of trainees and the objectives of media systems, the latter have to meet the following demands:

- The multimedia systems must be consistent, compact, easy to manage and "100 % compatible";

- The multimedia systems must include all forms of teaching and learning aids, such as machines and function models, hardware and software, readers, instructions for trainees and instructors as well as testing materials;

- Machines and function models must indicate typical, fundamental and exemplary criteria for industrial systems to enable a logical change from model level to reality;

- User interfaces of the control systems have to meet new ergonomic standards (e.g. SAA) and psycho-physiological parameters of learners.

(2) Description of the material flow

One can, for example, produce and code geometrical data with the help of a CAD program and convert them to a CNC simulation program. After the job has been entered, the actual goods in the store are indicated and a rough workpiece, transported on a shelf conveyor, is deposited at a delivery station.

The pick-and-place-robot takes the rough workpiece and puts it on the support of the apron conveyor. Now the workpiece runs to the colour-measuring system and than to the depth-measuring system. After the checking process the rough workpiece transported on the apron conveyor is deposited at a delivery station. The loading robot takes the workpiece and puts it into the chucking tool. There the rough workpiece is fastened and along with the chucking tool it is carried on under the CNC milling machine. Now the workpiece is shaped according to a DIN data set. After the shaping process has been finished the chucking tool moves to the loading position.

From there the loading robot takes the shaped workpiece to the work support of the apron conveyer. Now the workpiece runs under the vision system. After the geometric patterns have been checked, the finished product runs to the delivery station of the elevated shelf store model. Now the shelf conveyor takes the finished product, including pallet, and stores it in the elevated shelf store model. At the same time the goods in the store are updated. Now the output printer invoices the "costs". At this point the job has been completed.

(3) Costs

The costs for a complete CIM Learning Factory, including 12 traineeships, are approx. DM 600,000. The number of tenders for these complete systems are limited.

2.3.7 Experience and prospects

(1) Experience

At the moment experience with the complete course is available in a pilot study with 38 trainees. Excluding the automation technology level, a period of approx. 200 h can be assumed. The high amount of motivation and acceptance by learners has to be emphasised.

(2) Prospects

The prospects of such a model in outline form are:

- Extension of experience by training learners in the CIM Learning Factory;

- Commercialisation of the complete system of the CIM Learning Factory;

- New establishment and cultivation of partnerships between universities and enterprises;

- Optimisation of any part of the CIM Learning Factory;

2.3.8 Organisation development in enterprises

The contribution of this area has been described in point 2.3.4 on page 10. An essential advantage of the CIM Training Factory is that it makes it possible to offer training in the basic variations of work organisation.

3 Summary

In this paper a nearly perfected concept of basic training in the field of "Computer Integrated Manufacturing (CIM)" has been explained.

With the help of detailed studies conducted in part by the Department of Technology and Education, Department of Mechanical and Industrial Engineering, University of Dortmund the necessity of basic training at all levels for employees in Computer Integrated Manufacturing was verified. Then the new requirements for employees were indicated with respect to the "ability to act". Moreover, the didactic demands of the concept for basic subject-specific training were clearly stipulated. In summary, this concept has to include the invariant, indispensable, fundamental and exemplary contents and the basic options of CIM work organisation which are most important today and in the near future.

Then a configuration was presented to meet these demands: the multimedia system of the CIM Learning Factory, subsidised by the EC in the COMETT programme.

The CIM Learning Factory consists of

■ a well-operating "model factory", where activities like job management, production control, design, manufacturing, including loading, material transport and assembly as well as quality control and warehousing, are flexibly shown in functional models and are controlled by means of cross-linked computers (MPC); during the training the cross-linked computer structure is used like a language laboratory;

■ two different "teachware packages", the first for the target group of designers and decision-makers, the second for skilled workers and plant management. Both packages contain a reader and instructions for the trainees and an instruction manual for teachers/instructors. The latter book includes all the training media (transparencies, slides, video sequences).

4 Sources

/1/ Guggenberger, B.: Zwischen Ordnung und Chaos, Frankfurter Allgemeine Zeitung, 02.02.91

/2/ The amounts have been taken from: AWK (Ed.), Produktionstechnik, VDI-Verlag, Düsseldorf, 1987, p 536

/3/ Burgmer, M.: Gestaltung eines Aus- und Weiterbildungskonzeptes für die rechnerintegrierte Produktion (CIM) im Handwerk, Schwerpunkt Basisqualifizierung, expert, Dortmund, 1990

/4/ Institut für Sozialwissenschaftliche Forschung, ISF, München, 1990

CIM - Postgraduate Studies. Job-Integrated Training for Engineers

Prof. Peter Fröhlich / Prof. Dr. Hans-J. Holland
Department of Mechanical Engineering, Fachhochschule Wiesbaden
Wiesbaden Germany

Abstract

At the initiative of seven professors of the Department of Mechanical Engineering of the "Fachhochschule Wiesbaden" (Wiesbaden Polytechnic) the Ministry of Art and Science in Hesse (Germany) has established postgraduate studies. The aim of these postgraduate studies of CIM techniques is to impart knowledge with respect to computer applications in the field. of mechanical engineering. In the case of CIM the idea of integration rather than the single CAX techniques is emphasised in the course of the studies.

1   Background

The idea of establishing the postgraduate CIM studies was born out of the needs arising from the enormously increasing number of computer-aided applications in conjunction with a growing demand for adequately qualified employees. The requirements regarding employees' training (especially in the technical scope of work like design and development or production) are increasing rapidly.

Many industrial engineers only have fragmentary knowledge of the possibilities and problems which result from this evolution. More and more skilled people have to work with computers and software they did not become familiar with during their college or university education. Often there is only very limited knowledge of specific programs or single software products which has been picked up in crash courses or everyday work. Usually there is also no awareness of the possibilities and problems of integration and interfaces. Therefore, the computers and software are used improperly and very inadequately with poor results for the employees and the company. The considerations leading to the establishment of these postgraduate studies were initiated in view of these circumstances.

The offered postgraduate studies are based on knowledge gathered during previous studies and vocational experience. They point out the difficulties, possibilities and techniques of computer applications, but are not limited to furnishing special knowledge about specific hardware or software. Thus they provide all participants with the necessary "tools" to apply new computer technologies more advantageously in the future.

The postgraduate studies have been established at the Polytechnic because of the excellent preconditions for an effective curriculum, i.e. the good relations and close links to industry which are a prerequisite for effective and job-related courses. All professors involved have several years of industrial experience. They still maintain close contact to many companies where they were previously employed or which they became acquainted with in the course of common projects. In addition the assistant lecturers at the "Fachhochschule" mainly come from local companies, thus ensuring permanent input with respect to industrial experience.

Another source for feedback from "practice" is the engineers who were educated at the "Fachhochschule". They maintain their links, especially if they do not work for a big company and consequently do not get any scientific support. Cooperation through all these channels is both a necessity and a mission because among German universities and colleges particularly the Polytechnics are obliged to ensure a practice-oriented education for their students.

2   Starting Point

The situation at the universities and colleges, especially at the Polytechnics, is characterised by an extremely large number of students. Currently there are 7500 students studying at the Fachhochschule Wiesbaden with an increasing trend (as at nearly every Polytechnic in Germany). At the Department of Mechanical Engineering there are more than 1300 registered students, about 100 to 150 of whom graduate every year as engineers. These figures are three times the capacity the Polytechnic was originally designed for.

Although there is a tight situation concerning rooms and funds and a great burden on the scientific and non-scientific staff, the first idea for postgraduate studies came up in early 1989. There were three basic considerations:

Firstly, the "Fachhochschule" is obliged to provide further education and training. Secondly, there is a corresponding demand, especially on the part of medium-size enterprises, which are normally not very well-staffed or endowed with funds. Often they do not have their own capacities for continuing education (as in the case of big companies). And finally the new postgraduate studies with their advanced level will provide feedback to the normal studies of the Polytechnic and will thus help to maintain state-of-the-art education.

A general agreement to the proposed concept was reached by the Department of Mechanical Engineering and the State Ministry for Science and Art in September 1990, and planning and preliminary activities have been completed so that the first twenty students will be able to begin their studies in autumn 1992.

Who may take part in the postgraduate studies?

In accordance with above mentioned considerations, the postgraduate studies are aimed at mechanical engineers, process engineers, economic engineers or engineers in similar fields who are already employed. However, the first announcements for the studies show a great interest of engineers who have just graduated from our own Polytechnic and intend to attend such courses subsequently.

A third group of applicants are employees (mostly from production departments) who do not have an engineering degree, but do have sufficient knowledge to continue studying. These candidates may be admitted to the postgraduate study programme if they are able to give evidence of adequate professional training.

3 The Aims and Priorities of the Postgraduate Studies

The aim of the postgraduate studies is the application of computers in the areas of design and development, analysis, simulation, manufacturing, manufacturing requirements planning and general computer science. The contents of the studies may shortly be described as single "CIM components" and their joint operation:

- Computer Science
- CAE (computer-aided engineering)
- CAD (computer-aided design)
- MRP (manufacturing requirements planning)
- CAP (computer-aided planning)
- CAM (computer-aided manufacturing)
- CAQ (computer-aided quality control)

Projects and exercises using unique examples for different areas illustrate the integration of the computer and software applications. Besides the technical and operational aspects, the effects of these new technologies are discussed from the sociological, psychological, physiological and legal point of view as well as regarding the organisational provisions.

How to Take Part in Full-Time Studies?

The postgraduate studies require two semesters. As the majority of the designated engineers work full-time, a corresponding time frame has to be established which suits all parties involved.

One cannot accept the claim that most companies are not in a position to release important employees for a longer period of time. On the other hand, one cannot expect an engineer to risk termination of his employment for his studies. Moreover, this would not serve the purpose of the initiators. The relevance of the studies should be consolidated through reference to everyday work at the company. Any other concept of full-time studies would not be advisable.

Evening studies in addition to the normal job would be the other extreme. We all know very well that computer technologies are mostly found in those fields in which people are already exposed to high stress. Therefore, a large additional burden does not seem to be acceptable.

Taking into account the above mentioned circumstances, a special concept of full-time and part-time studies for the organisation of the postgraduate CIM studies has been developed. The part-time studies are based on lessons held two days a week during the normal semester: on Friday by receiving time off from work by the employer as well as on Saturday. In addition to this, there are two-week full-time block seminars, one at the beginning and one at the end of each semester.

4 Employment Aspects and Effects of Two-Semester, Job-Integrated Postgraduate Studies

The feasibility of the presented job-integrated educational concept depends on several requirements:

- The company concerned has to release qualified and serious engineers one day a week for a period of about nine months. Most of these people are urgently needed in everyday work!

- The student has to waive his free Saturday. Besides the two ambitious exercise days, there is some preparation and repetition to be done during his remaining leisure time.

- The time for the block seminars (altogether two weeks in one semester) has to be in the form of holiday time or release by the company or a combination of both.

The question of payment for the time released has to be negotiated between the employee and the company. By giving the candidates the opportunity to study, the company has already shown understanding for the necessity of the additional education. Subsequently a wage agreement with part or full payment seems to be justified. Alternatively a kind of flexible working-hour agreement may be reached to provide for adequate compensation for the days off.

Innovative companies not only invest in hardware and software, but also in the training of their employees. However, the effects of a part-time reduced engineering capacity should not be underestimated. So it is very important that both parties, the company (especially the direct superiors of the candidates, group leader and head of department) and the candidate, are convinced of the advantages and benefits of such further training and find a mutually agreeable solution to overcome the difficulties during the referred time period.

5 Structure and Organisation of the Postgraduate Studies

■   The studies are job-integrated. That means that the student is still mainly occupied with his permanent job at the company as before. At the same time the student is enrolled at the Wiesbaden Polytechnic.

■   The duration of the postgraduate studies is two semesters of 24 lessons per week each. This is the total of both the block seminars and the Friday and Saturday lessons, which is equivalent to full-time studies. It is considered to be the minimum scope for gaining the necessary knowledge in a suitable manner. Extended studies would have some advantages but are regarded as inappropriate because of the parallel burden of the studies and the job.

■   The schedule of the studies is as follows: each semester has a full week block seminar at the beginning and at the end. During the semester (17 calender weeks) there are all-day lessons and exercises on Fridays and Saturdays.

With this model the load on both the employer and the students is acceptable.

5.1   Teaching Contents and Didactic Concept

The contents of the studies concentrate on the possibilities and applications of current and future computer technologies in the field of mechanical engineering. Figure 1 shows a scheme for the structure of the studies, schedule and contents for the first term and Figure 2 the same for the second term.

The importance of the basic idea of integration has been taken into account in particular by the absence of rigid distribution of subjects within the block seminars. As far as possible the problems will be treated in their entirety. Even the subjects during semester lessons are split up to the extent necessary for teaching the fundamentals.

The exercises deal with model projects that show integration and data exchange. The individual subjects are not spread over the whole semester but are taught blockwise. In this way it is possible to obtain a sensible staggering and overlapping that corresponds to the handling of practical problems. For instance, CAD and CAE will be carried out in parallel and before CAM whilst MRP, as a clamp, will be integrated accordingly within the above.

The intention of the postgraduate studies is not to make the students perfect in some special computer application or software. Rather, they should understand the possibilities of the appropriate use of this technology for the various areas within their company. Therefore, it is more important, for example, to show the effects of CAD design for tool engineering and NC programming than to specialise a designer in a specific CAD system. Only a designer who has worked with an MRP system can assess the consequences of his design work for tool engineering and shop control.

As similar correlations are in evidence throughout a company, the CIM studies are above all supposed to improve understanding for the increasing number of overlapping effects of one's own work on other departments and staff.

5.2   Methods of Teaching and Learning

As already mentioned above, there are common full-day block seminars in the first and last week of each semester. The number of students is limited to twenty per semester. In this way all lessons can be performed as seminars.

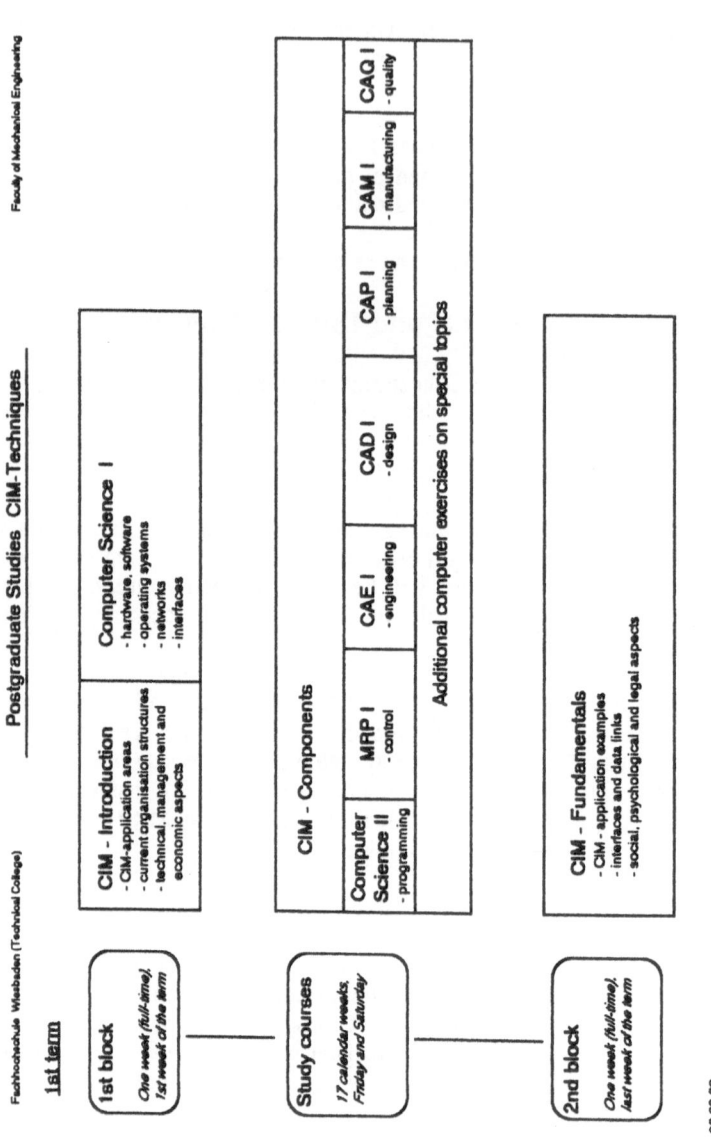

Fig. 1   Postgraduate Studies CIM Techniques, 1st Term

33

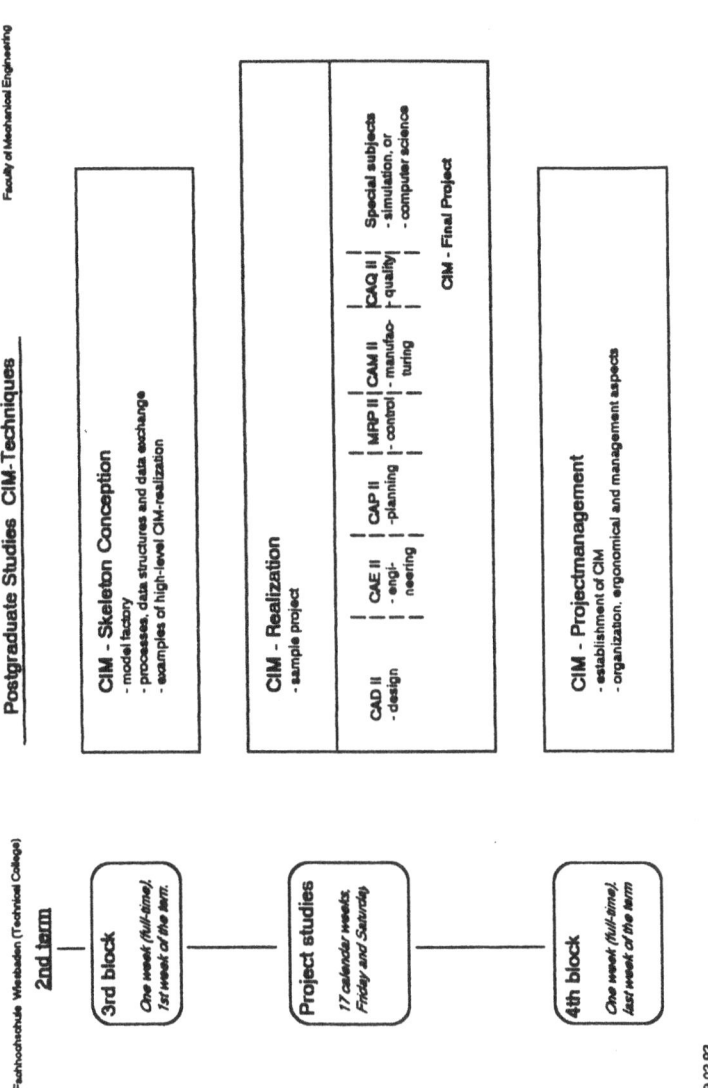

Fig. 2   Postgraduate Studies CIM Techniques, 2nd Term

Under the general coordination of a professor the subjects will be treated by various lecturers referring to their specific knowledge and background. As far as possible, the practical knowledge and experience of the participants will be taken into account and integrated. Demonstrations, computer exercises and excursions are provided for during the block seminars.

During the semester lessons both practical computer studies and seminars will be performed. Focus is placed on the practical work with the computer. The exercises will be conducted within small groups of ten students with two students normally working together at one computer station. All lessons are set up to establish close relations within a joint model project, which is handled during the course of one semester. On Saturday afternoons there will predominantly be guided exercises with the computers. During the second semester a final paper mainly dealing with the aspects of integration is to be submitted.

After passing the postgraduate studies, the students will receive a corresponding certificate. An additional report gives evidence of the individual results, which are to be elaborated in reports, lectures or test papers.

5.3   Equipment and Endowment

At the beginning of this paper the basic situation at the Polytechnic was briefly described as being characterised by an excessive number of students. It is, therefore, obvious that this project could not be started or performed with the available rooms, money and personnel. The normal courses of study already require a great number of assistant lecturers, additional space and funds.

Accordingly, a separate building converted for this specific use was rented for the postgraduate CIM studies. A number of seminar rooms, labs (equipped with workstation, host and personal computers) as well as offices for the lecturers, laboratory engineers and further personnel were furnished. This CIM centre is located near the Polytechnic, thus the regular students of the Department of Mechanical Engineering can use it as well from Monday through Thursday.

Two additional positions for professors, a laboratory engineer and a secretary, have been established for the new postgraduate studies . Further positions and the payment for the necessary assistant lecturers have been promised.

One of the most difficult and labour-intensive problems during the preparatory period has been the hardware and software evaluation. Only a few important criteria and requirements for selecting the equipment are summarised in this paper.

Starting with the contents of studies, these result in a number of necessary software requirements, which are covered by the commercial software market in very different ways. The software packages looked at were mainly CAD, CAM, FEM and simulation software; interface options to MRP software were also considered as very important. However, in spite of contrary information and statements of the suppliers, none of the existing software products is able to cover all CIM aspects satisfactorily.

The essential criteria for the final selection of the software were:

- Very good, easy-to-learn, self-describing and identical user interface in different program modules. There must not be much waste of time during the lessons to learn specific commands and procedures.

- Extensive performance in the field of mechanical engineering, especially regarding the CIM aspect.

- Suitable for the UNIX operating system (or MS-DOS for PC applications). Two aspects are decisive for this requirement: the lack of manpower (laboratory engineers) and the general trend in industry.

- Open interfaces and data link options with other programs, possibilities of data exchange, independence from hardware.

- Approved software products, widely spread and utilised in medium-size enterprises.

- Ability and willingness to cooperate on the part of the supplier.

- Favourable costs of acquisition and maintenance for the Polytechnic.

There is no fully consistent solution for all areas of CIM. The result is a partly heterogeneous software scenery with all its problems referring to hardware and interactions between the programs. Certainly the industry is faced with this current situation, too . Therefore, the equipment for the postgraduate studies is close to industrial practice, even though this is not a perfect solution. The advantage of heterogeneous equipment is that the students will be faced with all those typical integration and interface problems already during the training courses.

The hardware evaluation largely depends on the software to be processed. Regarding this aspect and the provided methods of teaching, the following basic configuration will be applied:

- central host computer with terminals for the non-graphic applications (like MRP or computer science)

- workstation cluster for the CPU-intensive and graphic applications (3D-CAD, FEM, simulation)

- PC cluster for MS-DOS application software (database, calculation, desktop publishing, text programs)

- various peripheral equipment (printer, plotter, data view, etc.)

Related to the requirements of the different courses the computers are equipped with differing memory capacities as well as computing and graphic performance.

Rapid development complicates hardware and software evaluation. The main criterion for the selection was not to buy the best solution, but rather to achieve the necessary and practicable things according to the aims of the studies. Top performance in computing and graphics are only necessary in some applications.

During the process of evaluation it has become obvious that all information and promises given by the suppliers regarding performance and short-term improvements must be taken with a grain of salt. As a general rule, one should only rely on performance and capabilities of systems that have been seen and approved by practical test. This approach, however, requires a huge effort when assessment of big software systems has to be done. External consultancy by industrial or college users may be helpful, though it cannot replace thorough knowledge of the subject.

Being aware of the importance of further education and training in computer science and application for their own business success, the software and hardware suppliers were willing to concede remarkable price reductions. Without these concessions the establishment of the described postgraduate studies in Rüsselsheim would not have been possible.

In spite of the massive support by the government of Hessen and by the hardware and software suppliers the postgraduate studies cannot be performed without charging tuition fees. Because of the high operating costs a fee of DM 3000 per semester is required for the postgraduate studies, but it can be reduced or waived according to personal circumstances.

6   Summary and Outlook

The initiators of the postgraduate studies in CIM technologies are aware of the fact that CIM is not a real training subject. The described concepts and contents cannot remain static. The present selection of topics is based on the establishment of the studies in the Department of Mechanical Engineering at a Polytechnic. In addition, the professors and assistant lecturers can include their own experiences and knowledge. Some interested people or readers may criticise the fact that some commercial or manufacturing aspects are not considered sufficiently, or that design, analysis and simulation aspects are too exaggerated. Depending on the personal and functional point of view it can be expected that social, ergonomic or management subjects have to be emphasised much more. All these topics will be involved due to the increasing application of computers and their effects on people, companies and the economy. However, the concept of such CIM postgraduate studies provides for a permanent adjustment of the intensity of its subjects according to current requirements.

Right from the beginning of this project, it was not intended to stick to a rigid concept. The Department of Mechanical Engineering believes it can provide a further training concept and program for the engineers of the region, especially for the employees of medium-size enterprises. They expect productive ideas from both the participants and the companies to maintain the up-to-date nature of these studies. The efficiency of the Polytechnic depends on intensive cooperation with business, associations, other colleges and institutions. Finally, the establishment of the postgraduate studies will be profitable for the regular students of the Polytechnic. This applies to for the additional technical equipment and the rooms as well as to the innovations that are initiated by such an institution.

Study Units in the CIM-Oriented Study Factory

Dr.-Ing. Martina Klocke
ÜAZ-Elmshorn
Elmshorm Germany

1   Basic situation

Nowadays more than ever, small and medium-size enterprises have to face the fact that the computer-integrated factory not only constitutes a future-oriented strategy of large enterprises, but will require, in the end, realisation concepts which are independent of the size of the business.

The information technique takes on special significance when the question is to respond to market requirements in view of flexibility, profitability, short process time, and reduced capital investment, and at the same time full use of capital-intensive operation material.

Here integration concepts have to be worked up for developed structures with the goal of linking existing "island solutions" and incorporating future-oriented system expansion.

New organisational forms here have to eliminate possible hindrances resulting from existing business management logistics and manufacturing technical process chains. Closely connected are new job structures and management systems, new job contents, and vocational descriptions which qualified and motivated personnel is faced with.

Employees being trained for modified methods of cooperative work need, however, new strategies for education, continuing vocational training, motivation, remuneration, and working hours. This is the only way to create the basis to fully utilise a continuous information flow.

It is, for obvious reasons, more difficult for small and medium-size companies, which constitute the majority of enterprises in Schleswig-Holstein, to organise systematic and far-sighted education and continuing vocational training work oriented to current requirements than it would be for large-scale enterprises. It is, therefore, very useful for them to obtain support from cooperating partners with qualified consultants for organisation, engineering, and education /1/.

With regard to local circumstances in the districts of Pinneberg and Steinburg, CIM-oriented aspects are linked to regional aspects and integrated in the education and continuing vocational training offered by the ÜAZ.

The inter-company training centre, called ÜAZ, is managed in the form of a holding association by uniting a number of firms, associations, and public institutions.

The two locations in Elmshorn and Itzehoe have experienced many years of practice-oriented work as an inter-company education and continuing vocational training centre for industrial and commercial firms in the districts of Pinneberg and Steinburg (Fig. 1).

The focal points set up now, especially for new technologies, such as EDP, CAD (computer-aided design), CNC (computer-aided manufacturing), SPS (stored program system), welding engineering, and packaging technology, are taught in education, training and retraining as well as in continuing vocational training (e.g. in certificate courses of the Chamber of Commerce). Presently, approx. 575 participants per day attend the daytime training courses, and 1.000 participants per year take part in evening training courses.

As shown in Figure 1, education and continuing vocational training participants come from very differing divisions of the enterprises. They essentially take responsibility for the reorientation of engineering and organisation in the enterprise. In practice today, the following demands are made on employees:

-       to grasp complex relationships and think in terms of systems;

-       to be able to filter out unnecessary information and retain what is essential;

-       to be able and willing to work in a team;

-       to be able to describe complicated circumstances in a comprehensible way;

-       to be able to formulate problems in a computer-compatible way and master new situations or difficulties.

Consequently, the education and continuing vocational training goals in the ÜAZ have to be set up in an integrated and interdisciplinary fashion. In order to acquire such key qualifications where professional, methodological, and social competence count, a Study Factory may be considered to be an ideal training facility.

The CIM-oriented Study Factory represents a technical-organisational copy of an entire CIM production plant; training components are implemented taking into account operational necessities and demands in the initial vocational training as well as in continuing vocational training.

The ÜAZ, which became involved in the operational use of computers for enterprise divisions, such as manufacturing, administration, design, and installation control, at an early date, has been developing a high-quality, flexible usable study location supported by the Schleswig-Holstein Ministry for Economy, Engineering, and Transport since 1989.

The project "Installation of a Study Factory taking into consideration CIM

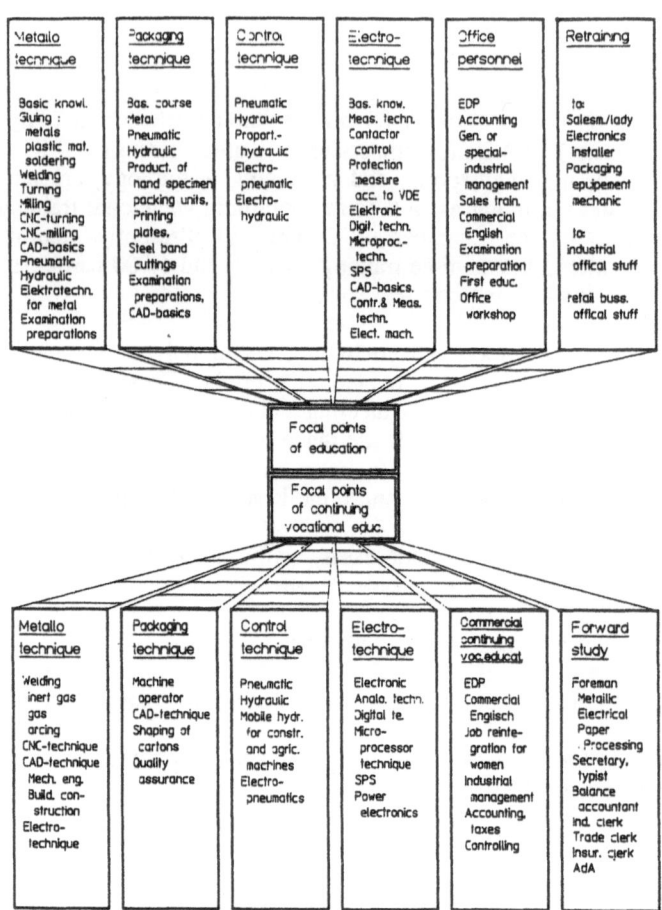

Fig. 1 Focal points of education and continuing vocational training in the ÜAZ

structures in the ÜAZ" is being carried out at the inter-company training centre by an interdisciplinary working team. This team is composed of division managers and instructors from commercial and technical divisions. In regular planning meetings the concepts available in the ÜAZ are revised step by step and introduced in an entity-oriented way. For this purpose, a product was developed and made available by the team, in which tasks were formulated, handled by individual participants, and examined with regard to their realisation.

Close cooperation is also maintained with the Fraunhofer-Institut für Arbeitswirtschaft und Organisation (IAO, Stuttgart, Institute for Work Economy and Organisation), with the Ingenieurgesellschaft für Arbeitswirtschaft mbH (IfA, Stuttgart, Engineering Society for Work Economy), as well as with the Institut für Pädagogik (Institute for Pedagogy) of the Christian-Albrecht University, Kiel. The ÜAZ informally cooperates with the Turmgasse Education Centre, too.

2 Target group

The CIM-oriented Study Factory as an integrated study location of the ÜAZ extends the field of conventional education, continuing vocational training, and retraining activities. With regard to the formal education goals and regular examination regulations, teaching content and curricula are constantly revised and adapted to reformulated goals of vocational training, as required for the reorientation of jobs in the metal and electrical sector as well as for office jobs.

With courses including innovative CIM aspects, the ÜAZ particularly addresses:

-       participants of education and continuing education courses for industrial, technical and office staff;

-       firms wishing to couple the introduction of computer integration and automation with organisation and training of employees;

-       instructors and employees having personnel responsibilities and wishing to become acquainted with new methods of teaching key qualifications.

The CIM aspects are taught from an entity-related point of view in the form of courses, training sessions and workshops as well as informally as instructor forums or informative presentations. Such meetings provide for continuous sensitisation in the enterprises.

3 Goals and subjects taught

The Study Factory pursues the goal of teaching operational sequences which result from the computer-integrated production through practical work. The CIM-oriented Study Factory in the ÜAZ represents a structurally developed medium-size factory manufacturing "rotary printing machines". The project team has developed the product rotary printing machine taking different aspects into account:

- By means of "standard implementation", the basics of the metal trade can be taught.
- The complexity can be extended by including further sectors (e.g. electrical technology, SPS, etc.)
- A close connection to packaging techniques can be established based on the nature of the machine.

The technical-organisational industrial process is, just as in practical operation, presented in the typical way of joint existence of fully and partially integrated computer-aided and conventional business sectors.

A great number of education, training & development sectors of the inter-company training centre are integrated in the teaching of work flow and organisational structures in the CIM-oriented Study Factory.

Practical education and continuing vocational training are no longer practiced in a specific "island solution". They are thus not only oriented to the operating material and the priority of imparting special competence, but are a living part of the integrated training "company". The participant will become aware of traditional limits of his special knowledge and obtain a transparent presentation of all operational events from an educational point of view (Fig. 2).

The participant is a "customer" and "employee" of the company at the same time. Upon ordering "his" rotary printing machine at the beginning of the course, he is able to follow up the total operational passage of his order, starting from order handling, design, project work and manufacturing up to assembly and delivery of the product.

Due to its complex structure, the rotary printing machine constitutes a product which allows the elaboration of interchangeable and multipurpose production modules as well as corresponding training modules. Especially the elaboration of "CIM-oriented training modules" has the purpose of meeting the demand for a flexible study location.

The variety of education and continuing vocational training courses shown in Fig. 1 is additionally broadened by the greatly varying length of the courses and differing qualifications of the participants. By developing flexible training modules, every participant should be able to get acquainted with the ideas of the CIM Study Factory.
The intensity of the participant's integration in the Study Factory is, however, dependent on his qualifications and length of the course attended.

4 Didactic information on course, training & project structure

Systematically projected further education is able to contribute to the motivation and flexibility of employees in the Study Factory as well as in operational practice. Demands regarding qualifications do change with reference to technical

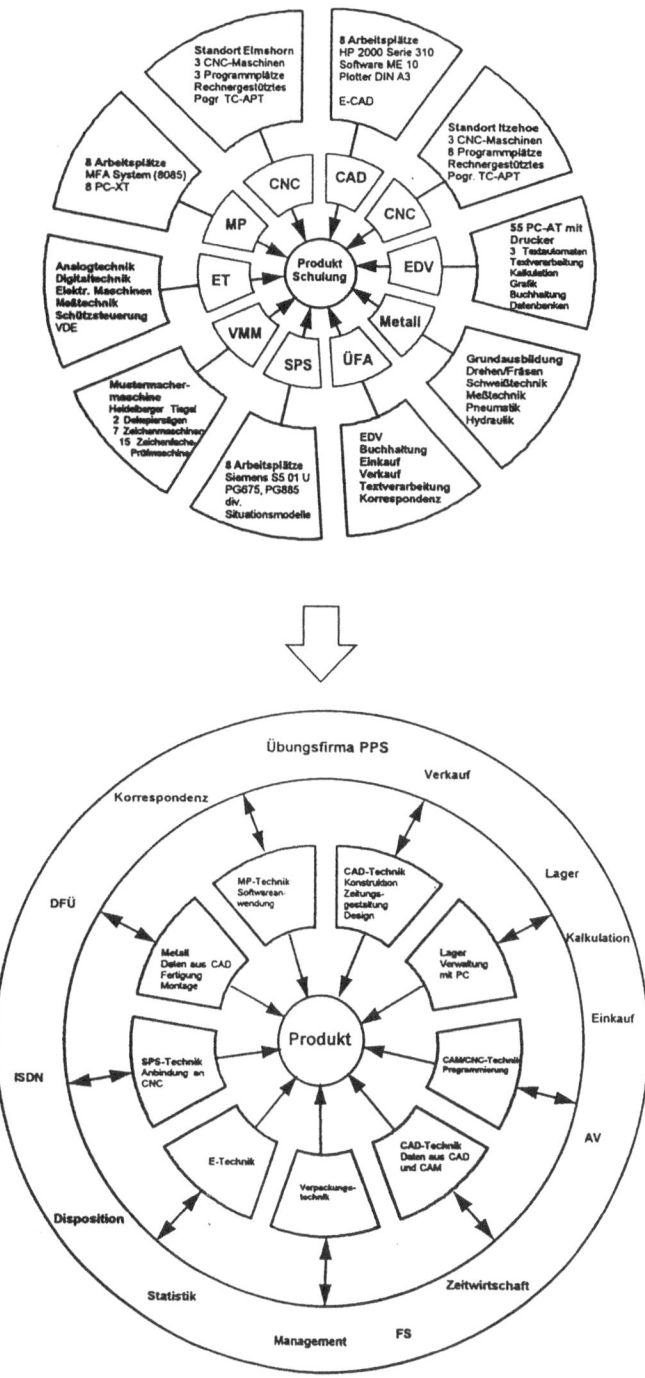

Fig. 2   Connection of the independent educational divisions to an entity operation

developments (Fig.4). Technology not only determines the type of education required; it develops with modified curricula and methods and is adapted to human-centered workplaces. Therefore, an organisation scheme for teaching and studying processes in the Study Factory has to be established flexibly. When developing modular teaching and learning subjects, the goals, subjects and forms of didactic organisation chosen have to be made concrete in the course of the respective education and further training activity. Essential characteristics of the CIM Study Factory are:

- Subject to learn is not the abstract topic or "a part for the scrap pile", but a real product (rotary printing machine) with the respective operational conditions.
- Operational reality is simulated as a model but is not idealised. The participant "goes through" CIM. There are real decision-making situations which will require thinking in linked, operational, and overall economic structures and may result in manifold possible solutions.
- Acting in the study firm means having to deal with data and structures as well as communication and cooperation in the group.
- Actions which in some cases become necessary under time pressure require strategies to solve operational problems.
- Activity is demanded of the participants, and they experience the consequences of their actions.

The most important objective of the project-related teaching and studying method is that the participant individually reaches the teaching objectives in spite of differing knowledge, understanding and memory, including checks of study progress.

Projects in this context are tasks given to participants who have to produce a ready-to-use product /3/. Depending on the field of operation of the participant, this could be finished workpieces as well as drawings, elaboration of geometric data, or an effective quotation.

Mutual dependencies must always be considered and the projects must be seen as an entity.

The subjects are divided into small study steps. The student is activated by tasks and questions. The instructor goes through the student's answers individually. The gradual changeover from simple to complex operational tasks means the individual student has to deal with functional divisions of the Study Factory as well as handling project tasks in a team /4/.

5   Teaching/learning organisation and methods

The circumstances of knowledge and learning of defined target groups influence the organisation of individual concepts for the course. Due to a certain flexibility of the organisation scheme, the instructor himself selects suitable curricula, methods, social organisation and media /5/.

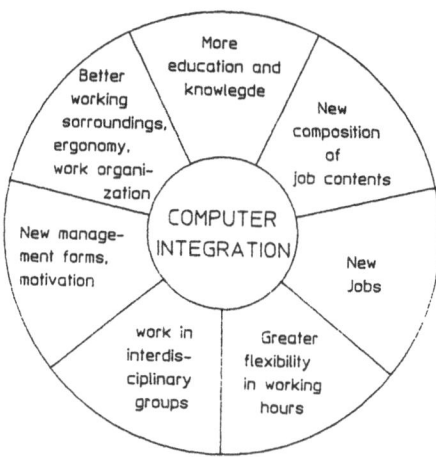

Fig. 3   Change in work due to CIM /2/

Free organisation of courses has the purpose of modifying modules, while taking into account the different educational circumstances of the participants. The aim is to develop different participant-related concepts, to analyse them with respect to their implementation, and to optimise them.

The scheme of a modular course in the Study Factory roughly looks like this /5/:

Start of the course

Orientation stage and theoretical introduction
to CIM in a large study group

Team organisation in small groups
Project work in the functional divisions of
the Study Factory

Team 1 Customer enquiry and completing quotation

-    Handling of form sheets and order forms
-    Determining expenditure in cooperation with design office, planning
     department, and quality assurance

- Determination of capacity utilisation as a project game, using plans, lists, and tabular calculations
- Material planning
- Elaborating surcharge calculation based on working programs
- Quotation (including word processing, method and terms of payment)
- Placing of order

Team 2 Planning department
- Production orders
- PPS planning
- Check: material stock - material requirement
- Determine material procurement terms
- Release material orders
- Release production orders
- Issue job control documents (e.g. check plans, drawings, work plans, parts lists, etc)

Team 3 Production/Manufacturing
- Incoming goods / quality control inspection
- Stock control via PPS
- Production (planning of details, and capacity reservation)
- Post-production calculation

Team 4 Design
- Modification of components
- Drawing filing system
- Time allowances
- Elaborate sample catalogue
- Explosion drawings
- Standard parts catalogue
- Assembly drawings
- Work flow drawings

Team 5 Assembly

Cooperation in a joint project (all divisions of the Study Factory participate) in the final large group review

The framework has a three-step modular structure /5/.

The first module serves to motivate participants and to provide the required study conditions for participants with regard to CIM. A survey is given of industrial processes, problems, and integration. Basic knowledge of other divisions comprises a description of the individual functions and components of the CIM module as well as their combined effect by means of the information flow.

The second module is to be regarded as a stage of application. It is divided into individual modules comprising partial divisions of the Study Factory. Project work in these partial divisions serves to employ group members in certain functional divisions and to familiarise them with the project work procedures.

In the third module, the problems of the individual job divisions are not of importance any more. Here, every employee of the divisions is supposed to use the linked systems in order to jointly carry out a total project. For this purpose, a customer order is handled completely. Starting from the product development until the selling procedure, all operational divisions have to participate.

The methods for teaching these subjects comprise a "mix of methods".
A "theory-practice" change takes place which has to be dismembered according to logical sections. In the first module, which has to show the cooperation of individual divisions and the concept of the Study Factory, methods include film visualisation of the subjects, metaplan engineering and instructional talks including foils and flipchart.

In the second module, which serves to support expert knowledge and key qualifications, project work is applied with guiding methods, such as guiding textbooks, instructions and questions as well as certain organisational forms like individual work, partner work, group work (Fig. 5).

In the third module, particularly thinking and taking action beyond the limits of one's own field, are required as well as working out strategies of problem solutions in the team. Depending on the length of the course and target group orientation, it is possible to handle each of the three mentioned modules or the individual partial sections of the second module independently from the others. As all the modules refer to the "rotary printing machine", any desired links of individual divisions are possible at any time.

The basis for this is a clearly structured integration of the study units in the total context in order to enable participants to always be oriented to the operational dependencies.

6 Configuration of the equipment

Due to the innovative commitment of the ÜAZ, firms, chambers and the government, new technologies were already able to be introduced in the inter-company education and continuing vocational training at a time when the reorientation in the metal and electrical trades was still in progress.

In this way , the planning of the Study Factory was able to be organised based on a high technical standard (Fig. 2). Figure 6 shows the original project as opposed to the actual project sequence. In the course of the second project year, robot technology was postponed in favour of quality assurance. This change in investment was decided in view of the demands of the enterprises. They have an increasing need for qualified personnel able to carry out jobs while maintaining and assuring quality and being familiar with the handling of increasingly complex measuring equipment and methods. This process includes all operational levels. Robot

48

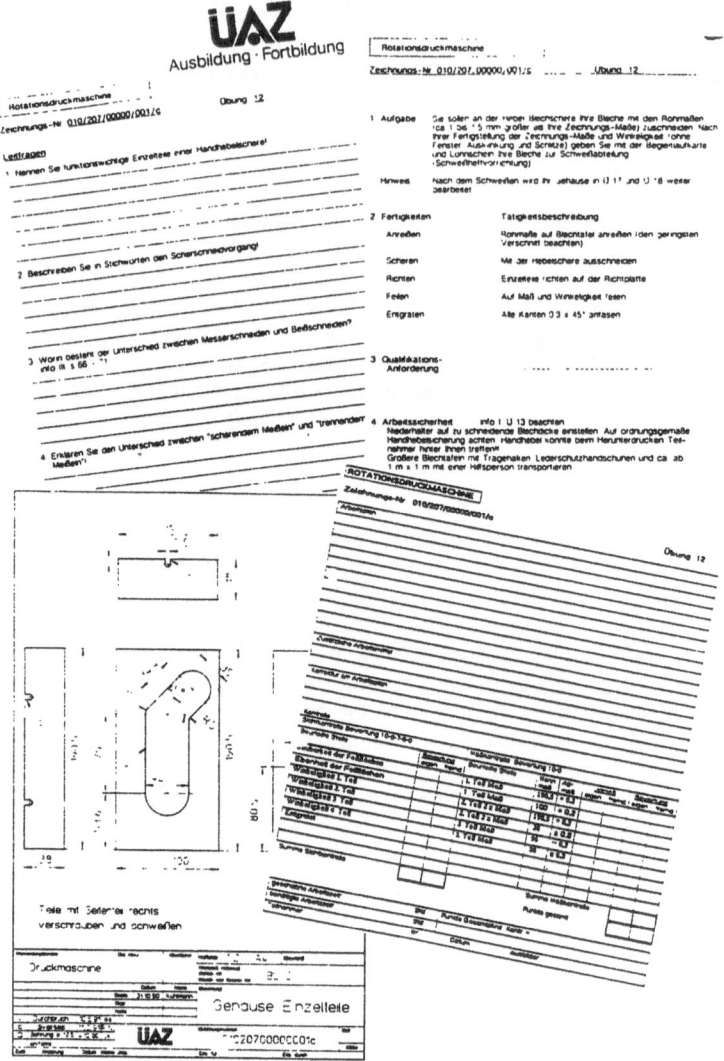

Fig. 4   Guiding text as a project work method

technology is at a standstill for the time being; human manpower is appreciated and accepted again as a production factor. Industrial robots have not succeeded in the region of Schleswig-Holstein with its small-business structure to such an extent that increased continuing vocational training is required.

Some of the existing "island solutions" in the ÜAZ were coupled.
In addition to divisions with information links via computer, in future conventional

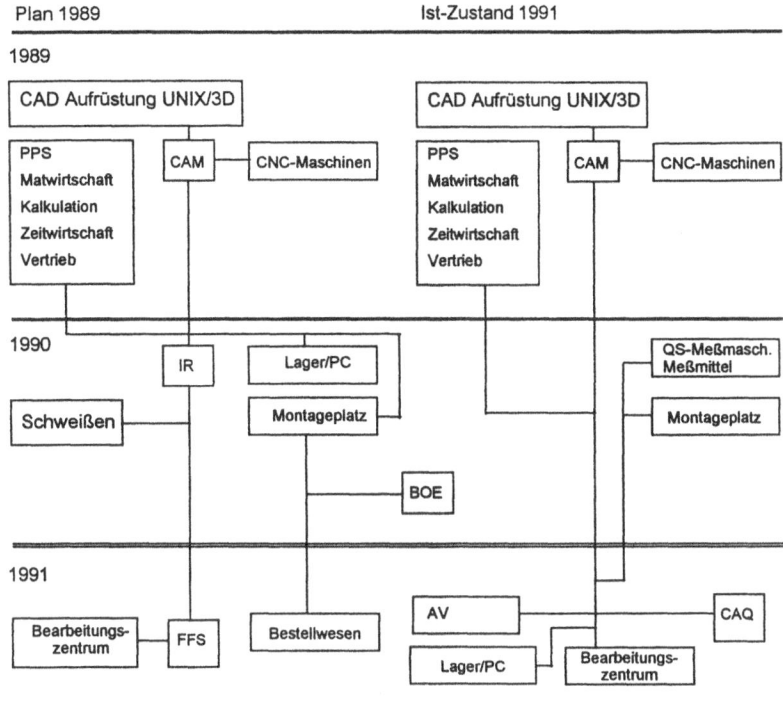

Fig. 5 Planning of the project "CIM-oriented Study Factory"

divisions will exist which will be integrated one after the other in the sequence structure of the Study Factory, including packaging technology as well as electrical and control engineering.

7 Experience, development and application prospects

Instructors:

Due to the freedom in selecting appropriate curricula, methods, social forms and media, high demands are made on the instructor. Different from didactic teaching, he/she has to have expert competence and knowledge in addition to a high degree of methodological and social competence. Furthermore, the instructor has to possess expert knowledge outside of his field, i.e. an engineer has to have basic commercial knowledge and an office employee basic knowledge in technical subjects. Self-controlled learning on the part of participants may lead to questions which cannot be answered merely through preparation of the subjects for training. The instructor has to change the method in time and determine by himself important focal points. That means he not only needs expert and pedagogical knowledge, but also has to be in a constant process of further education /5/.

Participants:

The most important problem regarding the participants is their heterogeneity as to education, age, capability to comprehend, willingness to learn, and their difficulties in learning. It is important to create as uniform learning conditions as possible at the beginning of a course. Brushing up basic knowledge is as important as making learning methods available.
Further, participants must be motivated because if they are ready to learn, a number of learning difficulties can be reduced or even avoided. As a rule, motivation is different for each participant. Participants sent to training "against their will" have to be motivated in a different way than participants who wish to broaden their knowledge voluntarily. The concept of the course has to be organised attractively and subjects going beyond the limits of the respective field must be chosen. The motivation is more effective the more it comes from the participant himself. The instructor can assist in pointing out the significance and practical fields of application.
Another problem is that of working in a group. There are students who have no experience in passing on their knowledge to others. They may obstruct work in the group by keeping their results secret. Participants who are not sufficiently motivated may, however, have difficulties in working on their own. A problem could also arise in the cooperation of technical with office divisions as the targets (e.g. quality and delivery date) cannot always be agreed upon /5/.

Background conditions:

The most important problem is the expensive equipment of a CIM Study Factory. Besides machines and computers involving high cost, corresponding software is also needed, and the study rooms must be well provided with media and training material. As the courses for continuing vocational training must be practice-related, machines, computers, software, etc. have to be as up-to-date as possible. Due to the investment to adapt the above, considerable unexpected cost may arise. Training material such as software and appertaining reference manuals is often not available yet. In order for the Study Factory to provide its own material, experts and considerable funds are required. In view of this cost aspect for many smaller institutes offering continuing vocational training, it will be impossible to offer CIM-oriented courses. In the future they will only be able to offer courses which are limited to the individual functional sectors (e.g. CAD for technical drawing personnel) /5/.

8 Transferability to other sectors

The most important objective of the Study Factory is to bring together technical and commercial divisions in order to support the firms and to prepare participants for cooperation in practice. The ÜAZ is increasingly forced by operational practice to develop new concepts for this, and individual branches stress the necessity to focus on the handling of their specific problems. It is, therefore, necessary to develop education and continuing vocational training modules not related to specific branches which thereafter may be complemented by branch-specific solutions.

Organisational, technical and personnel aspects have to be considered here as an inseparable unit.

Literature:

/1/   Klocke, M.: Fortbildungsanpassung durch Qualifizierungsberatung Management heute, 2/88, S. 11-14.

/2/   Spur, G.: CIM - Die informationstechnische Herausforderung. Vortrag anläßlich des produktionstechnischen Kolloquiums, Berlin 1986.

/3/   Klein, U.: PETRA Projekt und transferfähige Ausbildung, München 1990.

/4/   Behrens, A.: Ausserbetriebliche CIM-Qualifizierung. FgH-Berichte 2/88, S.63-66, Stuttgart 1988.

/5/   Utermann, D.: Entwicklung transferfähiger Elemente eines Curriculums zur Weiterbildüngsqualifizierung in einer CIM-orientierten Lernfabrik. Kiel, Christian-Albrecht-Universität, Dipl. Arb. im Fach Wirtschaftspädagogik, 1991

CIM Training in the Learning Factory

Siegfried Reith
CIM Learning Factory Turmgasse
Villingen, Germany

1   Development of our project

The project "CIM-Oriented Learning Factory" is a direct consequence of the industry-wide training which has been carried out and constantly further developed by the Winkler Company in Turmgasse in Villigen's Old Town. The training centre has approx. 260 participants per day in the following fields:

-       metal technology
-       wood processing
-       plastic processing
-       engineering drafting
-       industrial administration
-       production automation

providing training in "C technologies", such as

-       CNC technology
-       CAD/CAM technology
-       DNC/CAQ
-       PPS/operating data capturing/material data capturing

through the application of our CIM-oriented learning factory. In predominantly offering retraining, training and adapted continuing training programs in coordination with our Labor Office, we see ourselves as a partner of the craft trades and industry. In these sectors new technologies are placing higher and higher demands on the knowledge and skills of the work force. With our training, therefore, we want to stay one step ahead of the demand, particularly in small and medium-size enterprises. On the basis of the concept of a learning factory, we can cover the entire spectrum of industry requirements with practice-oriented continuing training. Our PPS-CAD and CNC (workshop-oriented programming - WOP) software, which is constantly kept up-to-date, makes it possible for us to come very close to achieving the goal of a CIM-oriented learning factory within the scope of daily training and continuing training. The Fraunhofer Institute (IAO) with its director, Prof. Dr. H. J. Bullinger, is furnishing the scientific support. When we began thinking about new content and ways of carrying out training work in early 1987, training on the CNC machine, design with CAD and use of the PC in office administration training had been more or less standard for years. Nevertheless, we saw the necessity of planning long-term further development (Fig. 1 - Schedule).

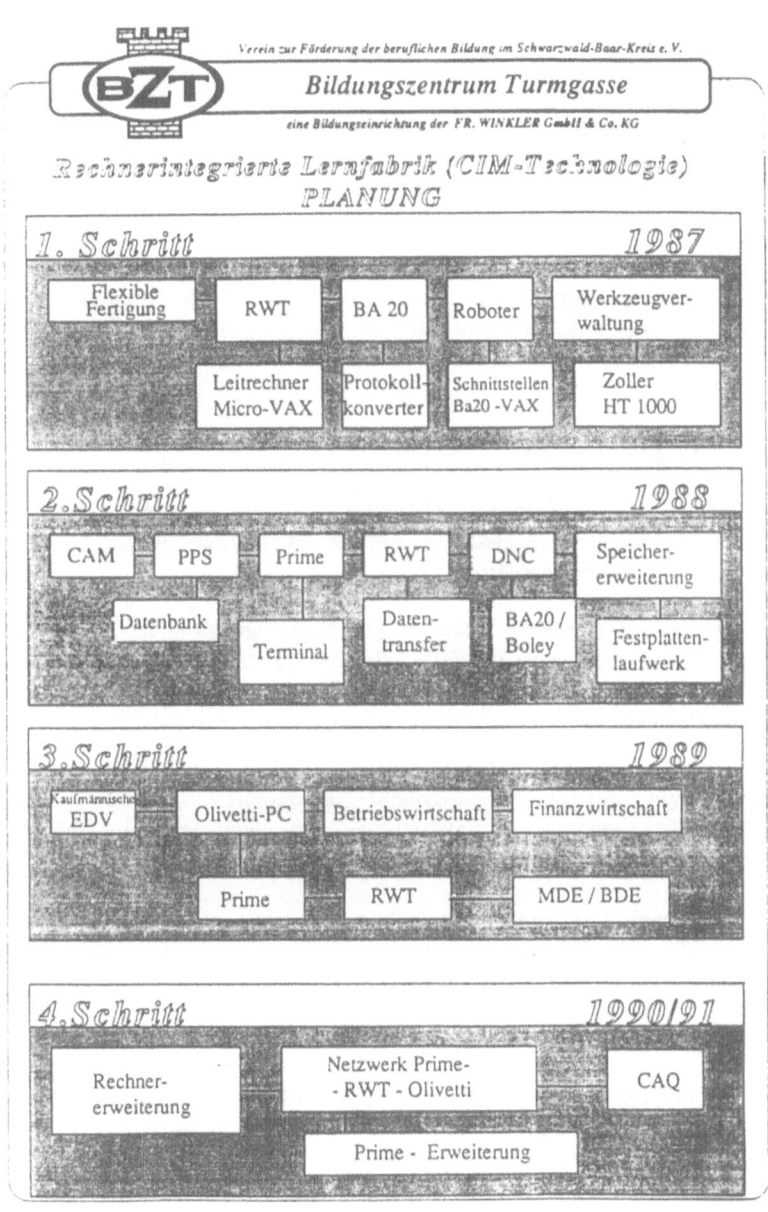

Fig. 1   Schedule

2   The problem to be tackled

CIM means, first and foremost, organisation and then meaningful utilisation of technology by qualified personnel.
Working in "CIM" requires more highly qualified commercial and technical staff members in all company departments. These qualifications can be obtained only in part in courses involving learning on computers, robots and NC machines. The abilities required for "interdepartmental thinking", the "ability to act in complex systems" can only develop if it is possible to gain direct experience with CIM systems.
The following, therefore, is based on the central thesis that future training and qualification needs resulting from new technology will not be determined by this technology so much as primarily by the new organisational forms which are reflected in CIM and are undoubtedly reinforced by technology.

3   Target group

At our training centre all those participating in long-term programs (6 months or more) are fundamentally regarded as "personnel". For the learning factory today this means a qualified reservoir of:

■   Production personnel (directly involved), approx. 110 "staff members" participating in technical-industrial training programmes in the following occupational groups: - industrial, design mechanic - carpenter - plastics moulder - metal-cutting worker

■   Production personnel (indirectly involved), approx. 90 staff members primarily taking part in the following programmes: - technician retraining, specifically in production   automation - draughtsman with 400 hours of CAD/CAM technology - system programmer/EDP technician

■   Administration personnel (commercial area), approx. 60 "staff members" taking courses at the "training company" and those doing retraining in industrial administration.

■   Executive staff members, who are basically the BZT instructors.

The above mentioned "work force" is also administratively handled as personnel in the learning factory (complete, simulated personnel administration). The staff is classified in accordance with the applicable wage agreement into hourly wage earners and salary earners. More importance is attached to the concept of treating training program participants as part of the work force than just providing "training material" for simulated personnel administration. "Personnel" is also an active element in a "CIM factory", for only when personnel acts, does a production operation take place at all. Thus, learning how to work in a "CIM factory" requires that a "work process" be carried out by the personnel. This is the essential aspect that distinguishes the "learning factory" from previous CIM training programs.

4 Aims/instructional content

The learning factory offers a complete, technical and organisational reproduction of an entire production plant corresponding to the state of the art of industrial technology. The result aimed at: creation of a high-quality and flexibly utilizable place of learning.

The learning factory pursues the goal of computer-integrated production. Since it functions as a training centre on an industry-wide basis, i.e. it has, in fact, developed just as "organically" as medium-size enterprises, conventional areas of work continue to exist within and alongside departments run using data processing. Thus the learner is not presented with a functioning computer world, as would be the case with purely computer simulation. Rather, he is faced with interface problems and frictional loss related to data processing in organisational and technical areas as well as in external relations which actually arise in practice. In this way such problems become perceptible and learnable.

The characteristic features of the learning factory as a place of learning will be outlined on the basis of examples: restrictions in the design of the equipment, for example, lead to less-than-optimum solutions in warehousing and in the material flow. Ideal high-tech ideas, such as computer-controlled shelf store, flexible transport system (FTS), etc., cannot be implemented by us at BZT. Numerous medium-size enterprises are also subject to such restrictions. In order to be able to utilise data processing meaningfully under these conditions, it is necessary to have a deeper understanding of the process to be controlled by means of data processing. Considerations of efficiency allow manual, hybrid and automatic systems to exist side by side in production. This causes considerable problems in production organisation. Decentralised data processing in the learning factory, such as CNC in the machining section, CAD of another supplier, our own EDP and PPS in the training company, results in system breaks which have and/or had to be solved technically and organisationally.

Working with such systems requires constructive, interdepartmental concepts on a basis of solid expertise and basic data processing knowledge. To acquire the latter, not only theory but also practice is necessary.

5 Didactic approach

Based on the instructional content striven for within the scope of computer integration, altered tasks with regard to the "integration path" result and from the wishes of small and medium-size enterprises come the initial contours regarding the learning goals. The latter, however, can be allocated to the individual functional areas only to a limited extent and are more oriented to the integration paths.

We have taken up the concept of "learning tasks" as our guiding principle and basis. A learning task is considered to be a field-related as well as a didactic unit of the imparting of knowledge oriented to the work tasks that actually arise or will in all likelihood arise in the integration paths as well as oriented to the learning goals considered to be essential for CIM:

■   Learning goals for the acquisition of specific occupational qualifications which can be determined from the vocational training regulations, the expectations

of small and medium-size enterprises and from the manufacturers' specifications.
■   Learning goals for the acquisition of data-related qualifications, such as system understanding, model concepts and ability to handle EDP configurations.
■   Learning goals for the acquisition of organisational and social qualifications, such as working in a team, planning, recognition of conflicts, decision-making preparation and systematic evaluation of information.
■   Learning goals for the acquisition of comprehensively cognitive qualifications, faculty of judgement, coping with complex situations, ability to be constantly involved in learning and the like.

The distinction here between knowledge and abilities is crucial since both make up a qualification.

6 Teaching and learning organisation

For CIM there are, as for other technical innovations, too, different strategies for dealing with expected training needs, such as

-        programs related to organisational setup
-        cooperative programs and
-        organisational training programs.

In most cases a combination of programs at various levels will have to be provided for the individual target groups. Particularly for a subject area of learning like CIM, which first of all emphasises very strong aspects of integral thinking and is, secondly, very company-specific in many respects, it is important to create a good link between learning and work activities.

The basic idea for such a concept of "learning and working" with the learning task is:

■   working in cooperative organisational structures. The learning task is organised in a cooperatively and integratively working group that is easy to deal with.

In this group, which consists of course participants in the areas of production, planning, design and administration with 3 persons in each area, there is a supply (customer order) of work-relevant information, especially computer systems. Time-related and field-related scope for action results.

■   Learning for the work: An organised form of learning with defined goals and content takes place.

CIM training is didactically and methodically prepared and is carried out without any work or production pressure. Learning and work contents are coordinated so that the knowledge learned suffices for execution of the job at hand (customer order) and can be utilised, maintained and expanded in the performance of the work.

■   Learning during the work: An organised evaluation of the work flows and work results takes place with respect to improvement of the working conditions and execution of the work as well as of organised learning.

During execution of the work an informal, cooperative training takes place via mutual assistance and instruction since all areas of the factory are represented in the "CIM team". In this way a high degree of specialised competence is achieved for all areas within the group. The role of the instructor is, to a great extent, restricted to the task of a moderator.

7   Configuration of the equipment

The learning factory at BZT possesses a computer-aided network of the entire order completion process with the following basic components:

- production planning and control systems (MODAS and PRODAT) in the area of commercial and technical planning tasks
- integration of CAD (PROREN 1 + 2 + E + NC and others) and CNC (RWT, WOP turning - milling - eroding) with joint access to stocks of master data (PPS database)
- actual data feedback (ODC/MDC)
- office technology with communications capability (networked Olivetti 486) and all administrative areas
- networking of the entire learning factory with all areas via LAN-DECNET for the complete order completion procedure.

CNC (metal-cutting) equipment is used in the production department. The penetration of technology will be limited to those areas in which it currently promises economic benefits in regular operations of the mechanical engineering sector. Assembly and installation, for example, is and will remain conventional.
An intentional criterion in the CIM-oriented learning factory is to provide for a so-called "mixed structure", i.e. highly automated, (flexible production centre with robot handling), partially automated and conventional areas must work together.

8   Prospects for development and application of training concept

In view of the relatively low level of requirements and experience for far-reaching CIM concepts in the majority of companies, the subject area of preparation of CIM training should be comprehended as an independent field of training and continuing training with respect to the factory of the future.
CIM training concepts require multipliers. The latter may, in some cases, also be books and specialist conferences, but people (instructors) are better suited because they provide for the implementation. Instructors are needed as multipliers for CIM continuing training concepts. Despite or, actually, because of the increasing complexity of technology, the importance of instructors must be emphasised. What is always required for a job in addition to specialised knowledge is knowledge and

skills obtained through experience ("having done it") and this can be achieved with our CIM concept.

9 Applicability/implementation concepts

The success of occupational continuing training programs in the field of CIM training depends to a great extent on whether the system knowledge related to the interconnection between and the integration of individual technical components can be imparted. Organisational qualifications not restricted to one field are of great significance in this regard.

The teaching and learning process must be organised along practice-oriented, prepared, integrated work flow lines. This also has consequences for material equipment and, in particular, staffing in continuing training programs and for trying out new forms of training and learning on site.

Planning for the introduction of technology must be linked more to the planning of organisation, personnel assignment and training than in the case of the use of individual computer-aided technologies. In this connection the training of planners and decision-makers themselves plays a special and new role.

10 Organisational development in the plant

In our view CIM is an integral concept. Any isolated or one-sided approach produces conflicts and disruptions in company work flows.

Particularly insufficient consideration of the persons involved, i.e. the staff who, in the end, must implement CIM, poses a threat for successful implementation. Therefore, it is necessary to redefine and redesign the interaction between man, organisation and technology within the scope of CIM implementation.

This new definition is also necessary because the character of organisation and technology changes as a result of CIM. Even the structural and work flow organisation has to be restructured. Moreover, greater competence for action can be given to the staff upon introduction of CIM, which may lead to a levelling out of the previous hierarchy structure. This would also be desirable in providing more transparency for organisational development.

The innovative idea involved in CIM is integration. This evolutionary process of integration between man, technology and organisation only functions (if at all) if one proceeds in easily comprehensible steps on the basis of clear, integral goals (as well as subgoals).

This step-by-step process of convergence has its own dynamics and requires constant regulation. The course must be continually compared to the goal and repeatedly corrected. In the discussion on CIM a distinction is made between two approaches:

-       an extremely technology-oriented approach,

-       a more human-centered approach.

The question of which approach will become the successful one depends on various

A laboratory learning factory called the "Rulers Factory"

Prof. Dr. Johan Vesterager
Institut of Produktion Management and Industrial Engineering,
Technical University of Denmark
Lyngby, Denmark

Abstract

This paper describes a learning factory called the Rulers Factory. It exists as a manual version, which already has been used for 10 years by vocational schools in Denmark, and in a CIM version, which for the time being is only a prototype. First the manual version is described with emphasis on technical aspects followed by an overview of different uses in vocational training. After that description of the current CIM version and a presentation of our experience in using this version are provided. Finally I will discuss the potential future applications of a CIM Rulers Factory.

1   Introduction

I would like to begin with a few opening reservations as to my qualifications in relation to this conference. I am pleased to have the opportunity to talk about our learning factory in this forum, but I am neither an expert in vocational training, nor an industrial psychologist, nor an industrial sociologist. I am an industrial engineer who studied the philosophy of management science in the 70s and has worked with new technology, especially in metalwork industries in the 80s, involving CIM or industrial information systems over the last 6 years.

As a consequence of this, my strategy in outlining this paper has been to stress the description of the configuration and the technical aspects of both the Manual and the CIM Rulers Factory in such a way that readers can hopefully imagine the broad range of applications of this laboratory factory for vocational training.

This does not mean that I will not discuss the current and potential future use of the Rulers Factory in its different versions. But I do not have an in-depth knowledge of all the present different uses of the already well-known Manual Rulers Factory at vocational schools in Denmark, and I do not have a precise picture of the future needs of vocational teachers as to CIM courses.

2   Outline

At first I will describe the Manual Rulers Factory, its history and its deployment today and then more thoroughly treat its current use. I have three reasons for doing this. Firstly, when teaching CIM by using a learning factory, it is excellent to have a manual version to compare with. Secondly, the CIM version contains, with only factors, such as product range, capital investment, CIM strategy, etc. According to our experience with training programs, CIM is a challenge to management and workers in companies alike. Those who strive for flexibility and a multifunctional way of thinking have to incorporate, at the same time, social competence and scope for action. Giving the worker more responsibility means trusting him more.

Summary

Technical innovation is not feasible without integrating man. This realisation must result in a learning effect for the CIM concept. It is important to show the course participant (user) as quickly as possible that the implementation of CIM concepts is possible with the technical means available today if man is integrated into the CIM process.

For CIM it is not modern data processing tools which are primarily necessary, the required organisational know-how is much more important. What is involved is a transition from work flow processes structured according to division of labor to integral work flows.

Such processes require workers who can think and act in a comprehensive framework going beyond one specific field. While great efforts are being made to enable the implementation of CIM on the hardware and software market, training and continuing training programs for integrated CIM concepts represent a bottleneck. Particularly the methodological and social ability of workers to deal with CIM is still greatly neglected today.

The focus of training is still on technical aspects, such as user training programs (CNC technology, CAD/CAM technology and the like). It has not yet been fully realised that CIM requires interdisciplinary ways of working (and corresponding authority and competence for taking action) on a permanent basis. CIM begins with rethinking and not with the purchase of technology.

The CIM-oriented learning factory can contribute to relieving the training bottleneck. Through the concept of the learning factory in Turmgasse we have created the comprehensive framework for reproducing in a realistic form the area of "the working world in computer-integrated, medium-size mechanical engineering enterprises" which has to be covered overall by training programs. Our CIM learning factory must not become a static model, just as a "real" company cannot remain in a certain state. It has to develop constantly and thus make new tasks and possibilities for vocational training available.

a few exceptions, the same equipment and information as the manual version so that going through the contents of the manual version makes it very easy to understand the CIM version. Thirdly, if someone should be interested in the CIM version, it would be absurd not to know all the possibilities which use of the factory in its manual version indeed offers for vocational training.

After the description of the manual version, I will present the CIM Rulers Factory, beginning with its origin followed by a detailed description of its configuration with special attention to its built-in flexibility. Finally I will talk about its use up to now and present some thoughts on its planned and potential uses.

3   The Manual Rulers Factory

The Manual Rulers Factory was developed around 1980 on the basis of a Master's thesis at the Institute of Production Management and Industrial Engineering of the Technical University of Denmark. The project was carried out in cooperation with the Danish company Bang & Olufsen. It has now been used in 10 years for the education or training of all qualification groups in industrial companies. Today there are around 60 Rulers Factories in Denmark, many used by technical schools (in training of skilled and unskilled workers), a few placed at the schools of the employers' associations, some used by consultant companies, etc.

The original intention behind the development was to teach blue-collar workers at Bang & Olufsen detail scheduling, group work, and the use of production specifications, all in preparation for a change in the work organisation (a work enrichment program).

As a laboratory facility or a learning factory, it is, of course, smaller and simpler than a real world production system. But still, it is a complete manufacturing facility with machinery and a substantial amount of paper work or information processing, i.e. the emphasis is on the planning, monitoring and control of a batch production system and not on sophisticated machining equipment. In this way the Manual Rulers Factory fully simulates a dynamic production environment.

3.1   The configuration of the Manual Rulers Factory

The Manual Rulers Factory exists in different versions. At first I will concentrate on the original version developed at the Institute of Production Management and Industrial Engineering. Figure 1 gives an overview of its components.

The figure is divided into two sections, one showing the direct labour area and the other the documentation reflecting the indirect labour. There are 7 process workstations for cutting (sawing), grinding, drilling, milling, painting, stamping and assembly. The machining equipment is composed of inexpensive manual electric tools, like manual drilling machines, pneumatically driven slides, etc.

Four different wooden rulers are produced in the shop. First the raw material (wood lists in two different widths) is cut into the right lengths for the batch in question. Next the ends of the ruler are sanded in the grinding machine to produce a nice finish. At the next station a hole is drilled in the ruler and afterwards chamfered - a one or two operation station depending on the tool used. At the milling machine one or two slots are cut, where later a rubber string (or cord) is put in at the final assembly station. At the painting station a 5 mm wide section of the ruler is painted

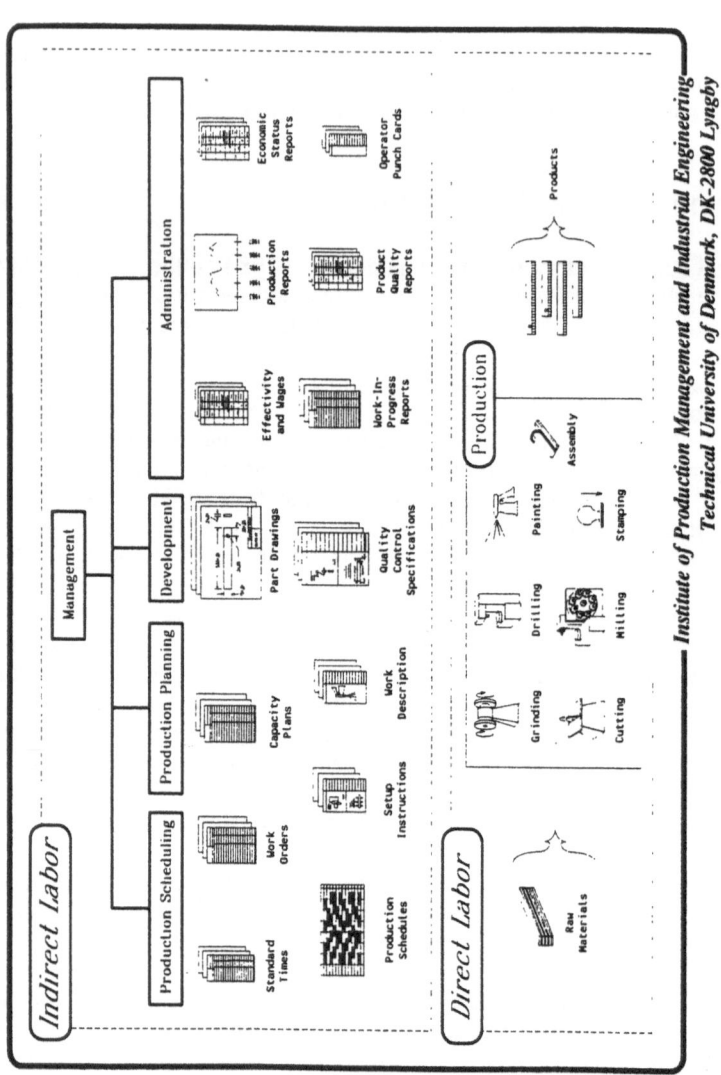

Fig. 1 Rulers Factory Manual Manufacturing System

as background for the subsequent stamping of the measuring scale. The final operation is then the assembly.

Not all rulers go through all operations. One is without a hole and rubber cord. Two have one rubber cord, the one also with a hole. The luxury ruler has two rubber cords and a hole. The rulers are also different as to length and width.

The area of indirect labour is concerned with all the documentation that specifies, initiates and monitors the production. It is divided into 4 categories: development, production planning, production scheduling and administration. Development consists of drawings and quality control specifications. In the original version there are only 4 types of rulers and thus there is no actual development during the operation of the factory - but of course, it can be done and is done otherwise.

The production planning documentation consists of set-up instructions, work descriptions and capacity plans (rough planning). Production scheduling involves standard times, work orders and the detail production schedule (a Gantt chart).

The administration papers include work-in-progress reports, product quality reports, operator "punch" cards, calculation of effectivity and wages, production reports, and finally, forms for making an economic status report for a production period (a calculation of the realised contribution).

The factory can be operated in different ways: with or without a foreman; with a fixed detail schedule from the beginning; with production orders which first have to be scheduled; with production of only 4 types of rulers; with an actual development of new designs as part of the exercise; with a sales department which competes with other factories running in parallel; with another salary system than the original designed, etc. The possibilities are almost endless, and many have been used at the schools in Denmark. Just to mention one more: I know of a course in material flow control or scheduling, where the same group first used the traditional Gantt chart for scheduling (a pushing system) and afterwards used a Canban pulling system in order to compare these different principles.

The Manual Rulers Factory has it own time (a large clock). In this way the teacher can stop production in order to discuss gained experience or make corrections.

The Danish company Grundfos in Bjerringbro has recently produced Ruler Factories in their own apprentice school. Theirs is an extended version comprising 13 operator stations beside the foreman. A new component in comparison to the original version is a saw for cutting the lists into 2 m pieces (so they can be machined in the planing machine), a planing machine (reduces thickness of the lists from 6 mm to 4 mm), 2 drilling machines instead of one, a milling machine for milling an ink edge, two stamping machines instead of one (a special one for stamping a logo). The price for this learning factory is Dkr. 86,241, excluding value added tax. Included are fittings and accessories, for example, 2 dust-removing vacuum units. In the last 5 years they have sold about 15 factories, including one to Sweden. Another company in Denmark has also sold Rulers Factories.

3.2 Use of the Manual Rulers Factory

As already mentioned the Manual Rulers Factory has been used to train all qualification groups, whether directors, engineers, foremen, skilled and unskilled workers, etc. Because the factory is easy to reconfigure, the training goals can be very different. The educational objectives include:

-        enabling students to experience the dynamic production world before attending courses in production management and industrial engineering;

- teaching the participants scheduling and different scheduling methods;

- teaching production specifications: how to develop and use them, the degree of detail in specifications, etc.;

- enabling the participants to experience group work with different salary systems;

- teaching different kinds of work organisation: autonomous group or foreman controlled group, the meaning of work enlargement or work enrichment, etc.;

- teaching quality control, the importance and implications of set-up and set-up-times, or different ways of improving working procedures.

Of course, the length of a course determines the number of topics dealt with. I know of courses where two or tree groups produce rulers in parallel and compete as to the best economic result. They manufacture rulers in 5 days with breaks where the single groups discuss how they can improve productivity, or the participants attend lectures on different work or productivity improvement methods, on the meaning and competitive importance of production flexibility, etc.

4   The CIM Rulers Factory

In 1987 we at the institute started a project called CIM/GEMS (GEMS = General Methods for Specific Solutions). The purpose of the project was to develop and transfer CIM development tools and methods to Danish companies in order to have their own production personnel understand CIM and develop their own CIM systems. The theoretical background of the project was the results from the US Airforce ICAM program, transferred to us by guest professor Robert E. Young (now North Carolina State University). The CIM engineering methods and tools were further developed, updated, and adapted for use by technical people who have little or no computer or programming experience. This to prove that CIM today is both affordable and accessible even by small companies. For a short description of goal, purpose, background, and rationale I refer to Vesterager et. al., 1989, for a description of results see Young and Vesterager, 1990. The focus of the project was on "the manufacturing of the manufacturing system" in contrast to the traditional industrial engineering "manufacturing of the product" viewpoint.

The project dealt with both technical and management issues of CIM, and the project had a unique project organisation which made simultaneous development, transfer, and testing of CIM engineering tools and methods possible. The building and the use of the CIM Rulers Factory had an important role in making this organisation possible. We used the Rulers Factory as our experimental laboratory for the development of CIM engineering tools and methods, as a demonstration facility in our transfer of knowledge, and as a template for our parallel industrial CIM projects. The CIM Rulers Factory is fully documented with regard to its use as a template. Of course our choice of the Rulers Factory as our laboratory was influenced by the fact that it was well known in Denmark and that most schools for vocational training and many other educational institutions already had the manual version. As a result, our development of the CIM version was influenced not only by present project needs, but also by the future possibility of later widespread application.

The requirements of importance in this context for the development of the CIM Rulers Factory were:

-   the factory should be low-cost and use only PC technology and, as far as possible, low cost off-the-shelf software

-   it should include both manual and automated operations because this is a typical configuration in small companies

-   it should be vertically expandable in the sense that it should be possible to add and manage more information and new stations without changing the basic system

-   it should allow customised production as well as being able to demonstrate other CIM benefits.

As a consequence, we developed a hybrid production system adding an inexpensive CNC engraving machine, but we did not go into the design of time critical tasks (real time control). We developed the computer system as a transaction oriented system with decoupled operator PCs. All information between functions is passed through transactions with the database. To put in another way: we conformed to the ANSI SPARC Three Scheme Architecture and SQL.

We decided to keep the functional and information contents of the original Rulers Factory unchanged with the exception of the engraving machine. This means that in the CIM version all paper information and manual information handling is computerised. It also means that the production schedule is loaded into the system before an exercise or production run begins.

4.1   The configuration of the CIM Rulers Factory

Figure 2 explains the system configuration.
The computer environment is as follows:

-   There is a PC at each operator station (IBM model 60 with an Intel 286 processor and 2 MB RAM). They share the same basic architecture running under DOS with the same application software. This to reduce system complexity.

-   The user interface is designed in a WIMP (Windows, Icons, Mouse, Pop up/Pull down) environment to have the most user friendly presentation of information. We use Microsoft Windows. This creates a software environment essentially independent of hardware. Microsoft Windows can run on all IBM-compatible computers, and it supports most printers, monitors, character sets and country configurations.

66

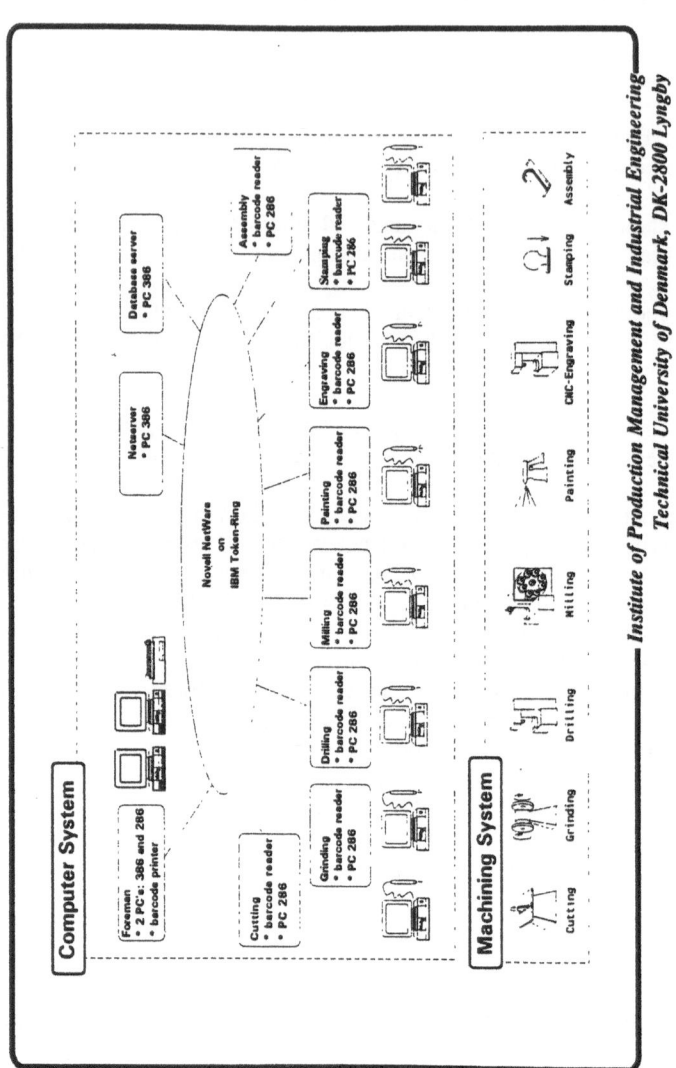

Fig. 2   CIM Rulers Factory

- All applications at the different operator stations have the same basic design. When logging in at a station, the software is downloaded from the fileserver and the database server. The applications are written in SQLWindows from Gupta Technologies.

- Operators retrieve information from and enter data into the system by using a bar code reader, which emulates the keyboard, and a bar code menu. The menu has codes for all the entries needed to perform the process operations (for the manual version: order information, part drawings, set-up instructions, work descriptions, etc., all displayed on operator request).

- The foreman station has a bar code printer and two PC's. This was necessary because of the 640k DOS limit. The bar code printer is a standard matrix printer. It prints the order identification code, later used by the operators to uniquely identify an order. One computer, equipped with a mouse (an IBM model 60), displays the Gantt chart, the order initiation application, and the clock application. The Gantt chart can be updated showing actual times for the start and finish of an operation. It is displayed below the planned times as hatched areas. The other computer, also equipped with a mouse (an IBM model 80 with a coprocessor), has the software for generating production results: work-in-process report, salary calculations, calculation of contribution, etc.

- The foreman software is written in Microsoft C using Windows Development Kit and Excel.

- All monitors are colour graphic with 640x480 resolution.

The local area network (LAN) consists of:

- an IBM Token Ring network with a Novell NetWare network operating system;

- a database server and a fileserver. In our system they are both 386 machines, though the database server could also be a 286 machine. The file server, which is the daily file server for the whole research group, has mirrored disks;

- database software called SQLBase from Gupta Technologies.

The production environment to process machines is the same as in the manual version, except the engraving machine. It engraves the measuring scale of the rulers and a name or a company logo. This allows for customised production.
The foreman's duties are the following:

- He initiates orders by printing out the unique bar code on a piece of paper - the only paper in the system.

- He monitors production by requesting an automatic update of the electronic Gantt chart. In the existing version the foreman cannot alter the initial schedule, but he can initiate orders at other times than first scheduled.

-       He accepts finished batches and updates the database accordingly.

The operator tasks are:

-       to receive a batch with the order identification bar code and to read the code in order to get the operation information on the monitor;

-       by using the bar code menu, to request the information he needs in order to perform the process operation;

-       to perform quality control and, by using the bar code menu, to update the database as to numbers of defect rulers;

-       to finish the order and, in accordance with the information received, transport the batch to the next station.

When the operator logs in and logs out at the operator station computer, the times are automatically logged by the database. When he starts and finishes an order, the times are also logged by the computer system. The last information is used to update the Gantt chart for actual operation times.

4.2   The use of the CIM Rulers Factory

The existing CIM Rulers Factory is still a prototype in the sense that it is not suitable for widespread use. This would require redesign in order to make it easy to install and written installation guidelines. Beside, some of the applications written in the 4th generation application generator SQLWindows have unacceptably long response times.
In addition, the existing version, a computerised version of the simple original manual factory, has no important manufacturing functions, such as an order reception function, a design function, a procurement and raw material stock control function, a detail scheduling function, a finished goods stock control, and a packing and delivery function - just to mention the most important. The existing CIM version has, nevertheless, already been used in courses and for the demonstration of CIM.

4.2.1   Use up to now

As already mentioned, we used the final CIM Rulers Factory in the CIM/GEMS project for different purposes. Important in this context is the emphasis we placed on CIM engineering tools and methods in the courses, such as the CIM project life cycle, technological, analytical and modelling tools to support work in different phases of the life cycle (function modelling, information modelling, prototyping, three scheme architecture, RDBMS technology, interface design, etc) as well as on the management side of the life cycle, including its relation to the technical side. In general, the participants first had to run the Manual Rulers Factory, and afterwards produce rulers in the CIM version. Depending on the situation, the teachers stressed different aspects of the CIM Rulers Factory or of its documentation.

The consultant group at the IPU Institute today offers courses in CIM for top managers, CIM project managers, and for engineers who have to deal with CIM requirements and design. All courses are based on our experience from the CIM/GEMS project, using the existing version of the CIM Rulers Factory.

The CIM Rulers Factory has been even more intensively used in a course at the Institute of Production Management and Industrial Engineering called "Industrial Information Systems". It is a graduate course teaching the students how to design CIM systems. A very important part of the course is group work, in which the students have to analyze, specify, make a prototype of, and fully document extensions to the existing version of the CIM Rulers Factory. In the last semester the students were divided into 6 groups of 6 students. The groups had to work together in twos. One group had to extend the factory with an order entry function, a rough capacity planning function, and a raw material stock and procurement function. The factory delivers four standard rulers from stock, but also produces, on demand, custom designed rulers with a logo. The other group had to make a finished goods stock control function which gives orders to the rough capacity planning function on standard rulers, a dispatching function which - depending on the delivery - was supposed to choose the best mean of transportation, and a function to receive customer inquiries on lost or late deliveries and trace the status of an order or delivery. The students, of course, had at their disposal all the documentation of the existing CIM Rulers Factory.

The course runs over 14 weeks with a 70-minute lecture on one weekday, and a 4-hour afternoon exercise, seminar or group work on the other weekday. Already on the first afternoon the students run the Manual Rulers Factory for three hours, and afterwards their experience is the subject of discussion. The next afternoon they are introduced to the CIM version and run the version for one hour. Again, the exercise is followed by a discussion on their experiences, and the teachers relates the introduced themes to the content of the course. See Appendix 1 for an overview of the course curriculum.

In addition to these uses of the CIM Rulers Factory, it has been utilised in demonstrating CIM for different university external interest groups. Just to mention some: the members of the technology board for the union of unskilled women in Denmark used two afternoons in first running the Manual Rulers Factory and afterwards the CIM Rulers Factory, followed by a discussion with the university research group on CIM and its implications. We also have arranged several evening introductions of CIM for industrial engineers using the CIM Rulers Factory as demonstration site. The joint technology board of the federation of metal industry employers and the union of skilled metal workers in Denmark spent an afternoon demonstrating the factory.

Moreover, the Rulers Factory in both versions is used for research purposes. At the time being a PhD student, working on the use and specification of object-oriented databases, is modelling the factory on the basis of object-oriented modelling techniques in order to study the object-oriented life cycle.

4.2.2   Planned and potential use

The CIM Rulers Factory concept has not yet been applied at vocational schools in Denmark for the training and teaching of skilled and unskilled workers. But, as mentioned, the factory has been demonstrated for representatives of relevant interest groups, and our consultant group at the Institute of Production Management and Industrial Engineering (IPU) is working on an application for raising money for an implementation project.

Therefore, I will discuss its different potential uses. The design of the CIM Rulers Factors, in accordance with Three Scheme Architecture, has made it flexible and expandable. There are, of course, limitations on the use of the existing version, especially for non-engineers because of excessively long response times for some of the operator application features and because of the limited number of manufacturing functions supported. Peculiarly enough, the limitation as to functions make it very useful in CIM design courses for engineers.

As a platform for the discussion of its potential uses, I will use the CIM project life cycle illustrated in Figure 3.

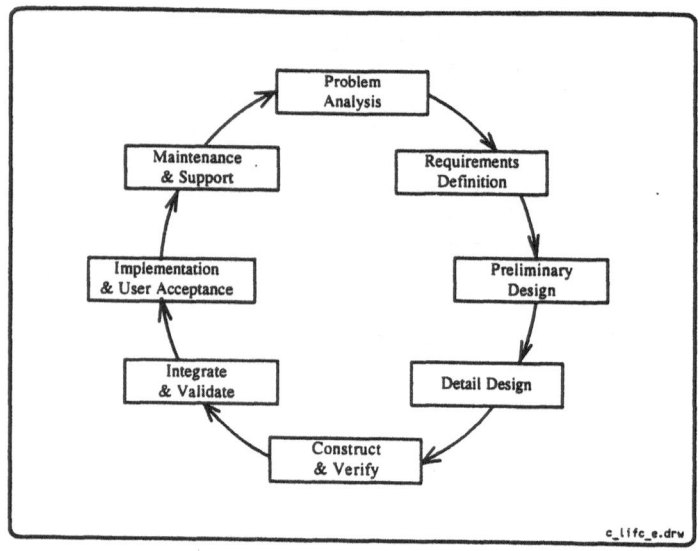

Fig. 3   CIM Project Lifecycle

The life cycle has three phases:

Phase 1:   Problem analysis and definition of requirements.

Phase 2:   Preliminary and detail design.

Phase 3:   Construction - verification and integration - validation.

Phase 4:   Implementation - user acceptance and maintenance - support.

The successful completion of a CIM project requires the participation of several persons in the industrial company, persons with different roles and qualifications. The most important participants are: the top manager, the project manager with responsibility for the daily progress of the CIM project, the domain expert who best knows the application area, the computer expert who knows the implementation technology, and the operator who uses the implemented system. Some of these persons might have overlapping functions in that someone may be both top manager and project manager, or both domain expert and operator. It all depends on the specific company and the domain. I will, therefore, discuss roles with specific qualifications, not persons.
In my view these roles should be linked to the following responsibilities (brief description):

Top manager:responsible for CIM strategy and formulation of strategic goals to follow in the specific CIM projects. Responsible for initiation of projects.

Project
manager: responsible for project organisation, project schedule, budgeting, control of progress, corrections, documentation, operator training, etc.

Domain
expert: responsible for the technical work in phase 1 of the life cycle. As far as possible, also responsible for phase 2 - the domain expert has to work together with the computer expert in phase 2. Together with the computer expert also responsible for phase 4.

Computer
expert: the computer expert is the phase 3 expert responsible for programming, network, etc. As for application construction, his role depends on the complexity of the applications.

Operator: responsible for commenting on systems requirements and design proposals and for daily use of the system.

The question is now: what qualifications do the different participants have to possess in order to ensure successful CIM project, and how can the CIM Rulers Factory and the Manual Rulers Factory together help in the training and education of these participants? I am convinced that practical experience is crucial for successful training of the participants. The apprentice model combined with crucial theoretical elements is the only tenable way in educating people in understanding

CIM. Just to mention one example: In our own course in "Industrial Information Systems" we asked the students if we could reduce efforts spent on the exercises in the factory and in group work in specifying new functions to the existing CIM Rulers Factory. The answer was an unambiguous NO!

Of course, I cannot give a full answer to all the questions on the potential use of the Rulers Factory in a short paper. But I can suggest partial answers. As to the training of domain experts and project managers, I know the answer in the way we use the Rulers Factory in our existing course "Industrial Information Systems". I must add that I do not think the course fully qualifies a participant to become a CIM project manager. The student also has to attend other courses, for example, a general course in project management and a course in organisation theory.

As far as operators are concerned - especially unskilled and skilled operators, whether blue-collar or white-collar - it is important to know about the nature of the life cycle: for example, when and how can I exert influence on a project? According to our experience from the CIM/GEMS project, blue-collar workers can easily learn to read and understand a function model (we used the IDEF0 modelling technique) of an existing manufacturing system. However, it is necessary to supplement the to-be function model with exploratory prototypes of the future system in order to get a dialogue. This dialogue is crucial for two related reasons. It is needed because the operators knows the existing system best (they are the domain experts of the existing system), and it is the best way to get operators involved in and get them to accept the new system.

In my view operators also have to know about the overall functioning of a CIM system, such as about databases, the interfacing of functions, the near real time sharing of information. Furthermore, operators have to understand the concepts of "batch-size-one" production and concurrent engineering.

An extended fully documented CIM Rulers Factory would, in my opinion, be ideal for these purposes. For example, a ruler is ideal for demonstrating parametric design and concurrent engineering.

Also, the different applications can be designed more or less intelligently. We already work with an intelligent foreman station. A PhD student has designed a prototype of an intelligent electronic Gantt chart, and our consultant group IPU is involved in an ESPRIT project called "Design, Development and Implementation of a Knowledge-Based Control Station" (ESPRIT 5161). The Munich company AHP is the primary contractor.

In depicting future possible applications of the CIM Rulers Factory, it is important to remember that the simultaneous use of both the manual version and different CIM versions offers exceptional potentials for designing courses with different aims.

5   Conclusion

The paper has described a Danish learning factory called the Rulers Factory. The Rulers Factory exists as both a manual and a CIM version. The Manual Rulers Factory already has been used by vocational training schools in Denmark for the last 10 years and has proven to be an excellent teaching aid for many different purposes.

The CIM version today exists as a prototype version. The paper discusses different possible uses of the CIM version and concludes that, due to the built-in flexibility and extendibility, the CIM version offers similar potential for vocational training in the area of CIM.

Appendix 1: Course Curriculum

NO	METHOD	TOPIC
1	Lecture	Introduction of the course.
2	Seminar/excercise	Seminar on information handling. Exercise: the Manual Rulers factory
3	Lecture	Overview CIM project lifecycle: phases, technical and managerial side.
4	Exercice/seminar	Exercise: the CIM Rulers Factory. Discussion of experiences.
5	Lecture	Function modelling (IDEF0).
6	Seminar/exercice	Seminar on IDEF0 and modelling exercises.
7	Lecture	Introduction of group work tasks.
8	Seminar/group work	Seminar on the technical side of the CIM lifecycle. Group work.
9	Lecture	Relational databases: introduction, overview. Introduction to SQL.
10	Seminar/exercise	RDBMS seminar and exercises.
11	Lecture	RDBMS: integrity, normalization.
12	Group work/seminar	Group work and discussion of group work progress.
13	Lecture	Information modelling (IDEF1): overview and terminology.
14	Seminar/exercise	Seminar on IDEF1 and exercises on information modelling.
15	Lecture	Information modelling and RDBMS-design.
16	Group work	Group work.
17	Lecture	Prototyping
18	Seminar/group work	Demonstration of a 4.-generation tool. Group work.
19	Lecture	The international standardisation work STEP.
20	Seminar/group work	Seminar on standard applications and selection criteria Group work.
21	Lecture	The design of user interface: criteria for good design.
22	Seminar/group work	Seminar on CASE-tools and CASE-demonstration. Group work.
23	Lecture	CIM-strategies.
24	Group work	Group work: report writing.
25	Lecture	The management side of the CIM lifecycle. CIM-organization.
26	Lecture	Future databases: object oriented technology.
27	Seminar/group work	Guest lecture by a director of information technology. Delivery of reports.
28	Seminar.	Group presentations of group works and evaluation by other groups.

Fig. 4   Course Curriculum

Literature

/1/ Vesterager, J. et. al. (1989); CIM/GEMS: General Methods for Specific Solutions in Computer-Integrated Manufacturing; CIM/GEMS document CG/T4, version 1.2, September 1989, 5. pages.

/2/ Young, R. E. and J. Vesterager (1990); An Approach to Implementing CIM in Small & Medium-size Companies; Proceedings of CIMCON '90, NIST Special Publications 785, Gaithersburg USA, May 1990, pp. 63-79.

Network for Operating Production Technology

Jörg Kluger, Norbert Meyer, Helmut Richter
BFZ, Essen
Essen, Germany

1 Technicians for Modern Production - Training as a Potential for Economic and Regional Restructuring

Skilled workers and technicians who are being trained today must be able to operate complex production equipment tomorrow. The consequences of malfunctions and errors often only become apparent in later production steps so that there are, of course, special demands on a highly skilled specialised technical competency at the shop floor level. In order to operate this equipment, skilled workers and technicians must not only have sound specialised knowledge and understand production systems as a whole, they must also be able to assess the consequences of a malfunction of an individual assembly group or component for the entire production process.

In order to eliminate malfunctions quickly, it may be necessary for skilled workers and technicians to involve colleagues from other departments. A social competency is required which goes far beyond the demands previously made on the "shop floor people". Furthermore, modern production workers and technicians are expected to familiarise themselves as independently as possible with new production conditions arising from technical and work organisation innovations.

They must also possess a methodological competency which enables them to learn in and for the future. This requires the ability for flexible and continuous further training for new fields of activity where the exact objectives and contents may be not b concrete at the present time. Finally, if one follows the discussion - particularly in Europe - on so-called "human-centered CIM systems" (see Cooley, 1987 and Corbett/Rasmussen/ Rauner, 1991 for an overview), it becomes clear that qualified modern production workers are increasingly being expected to participate in shaping their workplaces and work organisation actively (creative shaping competency).

The economic and technological restructuring process of the coal and steel regions of North Rhine Westphalia, and particularly in the Ruhr Area (see Bosch, H., 1992 for the consequences of this for the field of further training), is increasingly necessitating the type of technician and skilled worker described above (see Meyer et al., 1988). Thus, far-reaching changes and extensive prospects for creative shaping - particularly of the organisation of work - are emerging (see Rauner, 1988). Aspects such as the integration of previously separate departments and fields of activities, networking, the shaping and understanding of complex systems and questions of communication and cooperation are being accorded central significance.

The need to develop modern work structures throughout the hierarchies and the incorporation of technical components and systems necessitates continuous further development of education or training concepts in the fields of initial training, further training, advanced training and retraining. Parallel to this, training objectives, the contents of courses, media and concepts must continuously be updated and improved (see Feldmann, Kluger, Langenbeck).

For the technician at every level, the described development brings with it a need to be open to continuous processes of further education and training in one's working life. Therefore, the readiness and motivation for, largely, self-shaped learning processes and familiarisation with new fields of work must be developed and promoted (see Heidack, 1990 for a summary). At the same time, "learning throughout life" must not develop into a "lifetime sentence of learning". Rather, field and multi-field oriented education and training concepts which overcome the separation of theory and practice and technical field systematics in the narrow sense in favour of comprehensive vocational training addressing many fields of experience and subjects are to be developed.

Future-oriented vocational training must, as a contribution to personality development, consider the person as a whole. Accordingly, learning processes must be shaped jointly by learners, instructors and teaching staff. The field of production technology represents a broad application field of different techniques and forms of organisation, which suggests that there is little sense in using standardised training concepts and media. Open methodological approaches which open up room for creative shaping to instructors and teaching staff are preferable (see Klaßen et al., 1990).

2 The PTQ Pilot Project in the BFZ Essen

The Berufsförderungszentrum Essen e.V. (BFZ Essen - Job Promotion Centre) was founded in 1968 as a non-profit organisation. Among the members of the organisation are the Federal Labour Office, the Federal Ministry of Education and Science, the Ministry of Labour, Health and Welfare of the state of North Rhine Westphalia, the city of Essen, employer and employee organisations and the churches.

As a pilot institution for the vocational education and training of adults, the BFZ Essen is in the position to offer 1,200 retraining and advanced training positions in the fields of metalwork, electrical engineering, commerce, horticulture and landscape gardening, construction and its auxiliary trades as well as timber technology. In its function as a central point (ZÜF), the business dealings of about 650 training firms throughout the Federal Republic of Germany and in neighbouring countries are coordinated in the BFZ Essen. A corresponding pre-course support system serves about 15,000 participants a week with learning programs to prepare them for vocational retraining or advanced training. In addition, technical and didactic method seminars are offered to instructors and teaching staff as further training. Research projects and pilot experiments are among the innovative tasks of the BFZ Essen.

The BFZ Essen has established a modern advanced training centre in Chemnitz with 320 trainee positions in the fields of metalwork, electrical engineering and commerce in cooperation with the city of Chemnitz and with funds provided by the Federal Labour Office and the District President of Chemnitz. Since August 1990 a further point of emphasis of BFZ activities has been the training of teaching staff and instructors from the new federal states; this often involves advice and help in developing capacities for further education and training.

In the past two decades the close interlinking of scientific and "real" work with very different target groups has evolved concepts for vocational training and training which have enjoyed a high degree of acceptance and broad transformation both in the Federal Republic of Germany as well as in many European and other countries.
The pilot project "Production Technology in Regional Cooperations (PTQ)" (duration: 01.04.1989 - 31.12.1992; sponsored by: Federal Ministry of Education and Science and the state of North Rhine Westphalia) regards itself as a joint process of learning and development for all participants: together with an interdisciplinary project group (engineers, technology experts and technicians from the fields of electrical engineering and metalwork, teachers, psychologists and social scientists), more than 150 partners in the PTQ project (companies, vocational training schools, intercompany and external training institutions, manufacturing companies and research institutes) participate at different levels (BFZ Essen 1991, First Interim Report on the PTQ pilot project).

The fields of activities of the PTQ project in detail:

Development of training concepts and media

Regional and branch-specific training requirements are being elaborated jointly with the partners of the pilot experiment. Corresponding technical and didactic training concepts and media for the different target groups are being developed and tested.

Learning cooperatives and regional cooperation

In a current total of eight so-called regional groups, the different partners are developing common technical and didactic concepts and methodological procedures. At some sites the media developed have already been tested with different target groups in schools and companies. The experience gained from this form of regional and national cooperation is being used and tested on a broad basis in, for example, new regional groups in the new federal states and in European cooperation networks.

Joint further training of instructors and teaching staff

For the transformation of comprehensive action and project-oriented forms of learning, a joint basis is being created for the participating instructors and teaching staff. In addition to further technical training, this joint basis also takes particularly pedagogical, didactic and methodological aspects into consideration.

Experts, counsellors and advisory groups

The experience and results gained in the PTQ project and the work steps necessary for this are discussed and coordinated at regular meetings with experts and scientists (from the labour market, vocational training and occupation system and from manufacturers and media suppliers). The instructors, teaching staff, school headmasters and training heads discuss the transformation of production-related qualifications in advisory groups.

Scientific support

The PTQ pilot project dispensed with separate scientific support in favour of a close intertwining of the individual points of emphasis; this close intertwining is ensured in particular by the interdisciplinary composition of the project group and the intensive cooperation with experts (see above).

3   PTQ Activities

Regional Cooperation

Thus far eight regional learning cooperatives have been set up in the course of the project: Aachen, Bergisches Land, Bochum/Herne, Dortmund, Duisburg, Ennepe-Ruhr-Kreis, Essen and Gelsenkirchen/Münster. Initial steps for regional cooperation have been discussed and coordinated "on the spot" in these regional groups (e.g. information on individual activities, exchange of resources, know-how, etc., carrying out of joint activities).

To a large extent the individual groups determine their points of emphasis themselves. They are led by a regional coordinator (as a rule a school headmaster or a head of the vocational training department) and are advised and supported by two members of the PTQ project group. This support also serves the continuous transfer of information to the other regional groups and to the project group.

The points of emphasis of the regional groups in detail:

- Establishing contact and getting to know each other
- Information/communication
- Joint determination of points of emphasis for training
- Development and evaluation of suitable technical and didactic concepts for different target groups
- Exchange of experience, equipment, personnel, know-how
- Development and testing of projects involving other learning sites
- Establishing contact with other interested parties

The testing of joint projects in the field of further and advanced training, transfer within the existing regional sites and the setting up of new regional groups - e.g. in the new federal states - form the particular focal points for future regional group work.

The status of the technical and didactic concepts was documented and published at the beginning of 1991 (see BFZ Essen: Dokumentation der Technisch-didaktischen Konzeptionen der Modellversuchspartner, Essen). For the most part, the ongoing advising and support of activities occurs within the framework of the meetings of the regional coordinators and counselling of the individual regional groups.

Model testing and evaluation on the premises of the project partners (e.g. joint project weeks in schools and companies) and at the BFZ Essen (in the electrical engineering and metalwork departments) are being documented and published parallel to the technical and didactic development.

Training Research

Training research within the PTQ project focuses on the application of qualitative methods:

-       Continuous staging of visits to companies and of opportunities to exchange experiences for all the participants in the project (instructors, teaching staff, heads of training, school principals and personnel managers)

-       Documentation and advice with the technical and didactic concepts of the project partners

-       Execution of case studies of companies (guided interviews with personnel managers, heads of training, instructors, technicians and trainees)

-       Work place and company analysis (at the premises of the project partners as well as in other companies in North Rhine Westphalia and the rest of Germany)

-       Expert discussion rounds (as group discussions)

-       Recording the learning requirements of the partners

-       Documentation of the competencies and subject fields which, in the case of the different target groups, can be covered with the media which have been developed.

So far the field work has been carried out continuously within the scope of the different project activities (in particular within the framework of the regional group meetings). In a further step, corresponding offers are also to be organised nationally.

In the meantime several company case studies of project partners and other firms in the Federal Republic have been carried out and evaluated. In addition, the training requirements and offers of the different partners at the regional sites have been registered and evaluated with the aim of initiating, beyond the collaboration of to date, cooperation in the form of a supra-regional further training cooperative (help for self help). Initial action on this will provisionally begin in the autumn of 1991.

Pedagogical Concept and Didactic Approach

The pedagogical concept is a further development of approaches of action-oriented learning on the basis of theoretical action models (in particular Volpert's action regulation theory, 1985) and comprehensive and non-directive pedagogical guidelines (Rogers, 1972, 1982, Tausch and Tausch, 1979, Hinte, 1990) and represents an attempt to connect these ideas with a systemic orientation.

The working person is regarded as a member of an open socio-technical production system who finds himself in a process of continuous development in which he is subject to general system mechanisms and rules (which have to be specified more exactly for each particular field).

The optimisation of the input-output correlations and improvements in working conditions, the relationships between employees, fields and departments and technical organisation present themselves as normative guidelines for our work. We do not understand the system levels or subsystems which come into consideration in each case as a priori realities but as functional, task- and problem-oriented arrangements. Naturally, prevailing structures (e.g. company structure, work organisation, etc.) are to be considered. These are regarded as being more or less fixed, although they are, in general, fundamentally alterable. Through them shaping approaches are given meaning and space.

The didactic approach aims at developing competency to act in a real job situation which expresses itself in flexible and creative problem-solving in complex production systems. Besides teaching technical competency, this necessitates the promotion of closely intertwined methodological and social competency in the overall context of the system (see Hellwig/Richter/Tepper, 1991). These different aspects of competency to act are experienced and understood in their interaction and in relationships in their system. It is, therefore, necessary to shape the learning environment and media so that learners are provided with the relevant system references and structures. In processing a task, the work processes and organisational structures should be as realistic as possible. In addition, the technical media must show relevant, realistic system structures in their organisation and be equipped at every system level (which comes into consideration) with industry-specific functional units, modules or components.

This is connected with the problem of system-oriented elaboration of technical contents. In its upper section the system scale comprises general company processes and integrates a view on production in terms of processes and subjects, and in its lower section it contains specific and detailed knowledge of the subject. In this sense, structuring of the levels of knowledge and independent elaboration and integration of new technical contents are among the central pedagogical teaching objectives. This involves, among other things, the ability to select a suitable system level for the description of functional correlations and process stages, even in communicative contact with colleagues from other technical fields. Thus specific technical knowledge and knowledge of other technical fields is required which has to be acquired and integrated in an overall store of knowledge. The acquisition of knowledge and the exchange of information are supported by these principles and the work procedure and experience of the learners are examined against this background.

In connection with system development, we regard the ability for continual self-training within the work process as a further training objective. Therefore, we place special importance on self-control of learning processes. Information materials are structured as realistically as possible. Instead of didactic reductions in the information supply, a parallel process by the instructor is dispensed with. He encourages the learners to develop their own approach towards the acquisition of contents and the systematic structure. A new role for the instructor can be derived from this requirement: the instructor is additionally challenged as a moderator, learning advisor and counsellor, ensures that there are suitable learning environments and appropriate media, helps in the diagnosis and treatment of barriers to learning, moderates reflection rounds, promotes the overcoming of learning obstacles and conflicts in the team, etc. Naturally a high technical competency is still required, but the instructor need not master all the contents; rather, it is almost desirable for him to be integrated into the learning processes as a member of the learning system.

The objective of this approach, which we call the participation principle, is that the instructor, who as a "competent learner" can teach the self-controlled acquisition of knowledge and skills convincingly and effectively, has a model influence. The largely self-controlled group process and training within a work team is described in detail and differentiated in the concept of cooperative self-training in Heidack (1989, P. 16ff). We apply this approach as a significant component of our pedagogical model, particularly with the inclusion of systematic aspects.

Adequate procedures to meet the need arising from a system-oriented viewpoint for a realistic representation of company structures and work processes can be found in the planning game method or training company model. For the field of production and technology a realistic situation is to be structured for handling relevant technical installations and machines. In this field the qualified work is not restricted to planning, data management, and information process but it is characterised by the integration of these aspects into the different steps of procedure.

Our concept of system-oriented action learning represents an embedding of action-oriented learning in a systematic context. The basis and starting point for this is a complete action model with the phases of diagnosis, information, planning, execution, evaluation and reflection.

In this model it is not so much individual actions which are emphasised but sequences of action and their hierarchical organisation. The qualified worker must have those structures to execute effectively the activities in a complex context.

A procedure practising the planning and the division of labour in a team can be considered from analogous points of view. The guideline for us is a (schematised and idealised) organisation structure:

Firstly, the individual tasks to be carried out comprise complete actions. Secondly, every colleague is - at a higher system level - involved in the planning and evaluation processes relevant for his activity. Thirdly, the entire group undertakes the planning and evaluation steps at the highest organisational level (where the entire action organisation is regulated). We call this case a complete cooperative action.

For the teaching of system-oriented action competency, we have developed a phased procedure which, step by step, includes further parts of the system. The entire production system is regarded as the focused point right from the start, and individual subsystems are taken and placed at the centre of pedagogical consideration without losing sight of the overall context of the system. In every learning sequence the learners (teachers and instructors as well) are dealing with expanded system correlations in which the previous subsystems are embedded (see the section: Advanced Training for Instructors and Teaching Staff.)

For our purposes, the form of project-oriented learning seems to be particularly suitable as a basis for the methodological procedure in the individual phases. In applying it, the previous central system aspect is integrated in a new and expanded field in every new case of project work.

Technical and Didactic Concept

Production workers and technicians being within education and training processes will later have to operate complex production equipment which is integrated in different concepts of work organisation, technical and company organisation developments. A series of practice-oriented, effective and "low-cost" learning media and materials suitable for training purposes have to be developed for the learning and teaching of competency to act in a real job situation. The design of learning materials has to meet technical aspects as well as social learning, methodological aspects and competencies to shape the design of the workplace (see above).

This competency to act in a real job situation (both for commissioning, operation and retooling for other products and processes as well as for maintenance and repair) must be taught on realistic production units which have been set up for training primarily in terms of didactic aspects.

The assembly plant illustrated below is structured in modular form offering a great deal of flexibility and room for shaping and modification while taking into account different training course contents, target groups and production processes. It consists partly of industrial components and partly of self-developed assembly groups that can be realised within the framework of training. The individual components, assembly groups, modules and stations can be regarded as independent systems in themselves and as subsystems which can be selected from the overall range of equipment available according to the principle of applicability, degree of flexibility, frequency and their exemplary character.

Within the framework of the PTQ project, the BFZ Essen is developing a plant which fulfills these requirements. The plant has been conceived as an assembly plant for hinges and consists of a total of four stations:

At the first station, a robot places punched hinge flaps on a workpiece support; at this station the programming of robots is to be learnt exemplarily.

The workpiece supports are then transported on twin-belt conveyors to the second station, the assembly station, where a pin is pressed in the eyes of the two hinge flaps. In addition to the basics of pneumatics, problems arising from the accurate feeding of small parts are demonstrated at this station. In this case, the hinge pins must be transported from a spiral conveyor via a linear conveyor and a singles marshaller to the hinge.

The workpiece supports with the assembled hinges then move to the third station, a testing station, which uses a video system to check that the hinge flaps have been properly assembled. The result of this check is stored on a coding memory mounted on the workpiece supports.

A portal loader sorts the hinges into good and reject pieces at the fourth station of the assembly plant. This portal loader has two horizontal arbors, which are driven via threaded spindles, and a vertical arbor in the form of a pneumatic cylinder. The travel of the horizontal arbors is restricted by limit switches, with the result that trainees have an alternative to robot control at this station.

Social and methodological competencies cannot be learnt on production equipment alone. Rather, it is necessary during training to create a framework within which the training course participant can deal with other departments and in which he can acquire new knowledge independently. Therefore, the PTQ project group offers a series of advanced training courses in which participants from vocational schools and training companies work on the already completed parts of the assembly plant under realistic conditions.

The realistic conditions are simulated in the form of a training company. While searching for errors or executing tasks on changes in design, mechanical and control functions, etc., the learners are compelled to initiate technical discussions with colleagues from other fields (electrical engineering/metalwork) and to involve other "company departments".

The control system of the plant was conceived such that it can be expanded and structured from a subordinate to a complex superordinate control system (different possible control concepts like master-slave system, network, etc.). The individual stations (part-insert station, assembly station, testing station, product removal station) can also operate independently (as stand-alone systems) with regard to the rest of the plant. Therefore, every station has its own autonomous control system.

ASSEMBLY PLANT

This design of the control system ensures that interferences or malfunctions of one of the stations cannot directly impair the others. The modular structure of the plant,

84

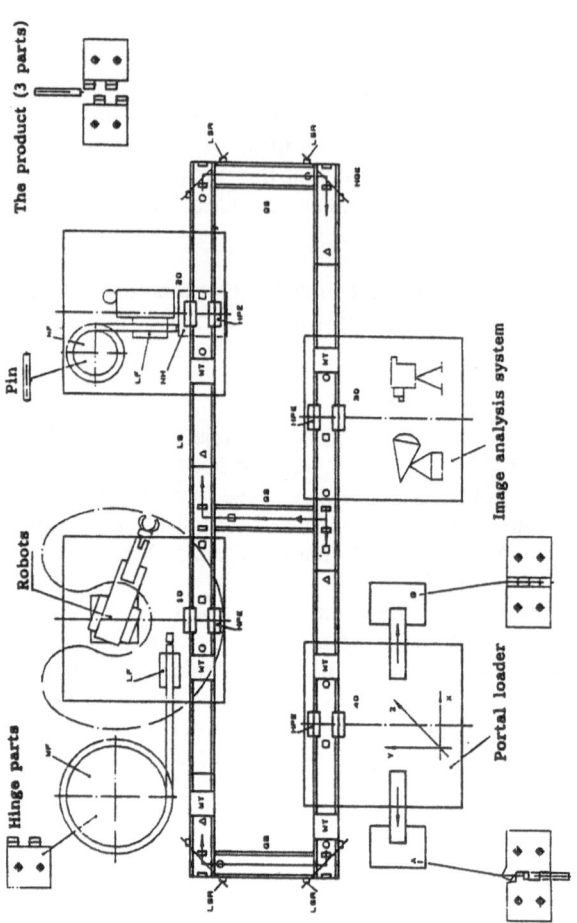

Fig. 1   Assembly Plant

which was one of the most important development preconditions, permits an immediate flexible integration into training of individual components of the plant. The complexity can be increased step by step and the correlations of the elements of the production system can be made visible (learning via progressive technical-organisational effort and methodological transformation.

The significance of the learners' work and activities performed become clear immediately. This fact additionally promotes the acceptance and involvement of the learners. At the same time, a phased familiarisation for instructors and teaching staff is possible, both within the scope of training/lesson preparation as well as in the learning process itself (e.g. with project work).

Components of the assembly plant have been used in former PTQ advanced training courses for instructors and teachers. It is planned to install further components and modules of the complex assembly plant with a corresponding transport system and to use them in the next series of seminars. In the further course of the project, the complete assembly plant - with corresponding technical and didactic materials - will be completed and tested and evaluated with different target groups.

Advanced Training Seminars for Instructors and Teaching Staff

Our pedagogical, didactic and technical concepts are transformed in a series of advanced training seminars for instructors and teachers. Three series of seminars have taken place so far (each with five parallel events for about 15 participants).

The themes of the first seminar were fundamental questions on production concepts, work organisation, technical development, team work and qualifications, the interaction of these different aspects and different learning and teaching philosophies. The seminar took the form of a "future workshop" (in accordance with Jungk and Müllert, 1981), supplemented with a smaller project work (application of control technology).

The future workshop is a structured form of exchange and creative shaping in changing group sizes (large and small). It is a creative shaping and discovery method which leads to definite courses of action in a final realisation phase. The project work represented an expansion by us and aimed at making first realisation steps as real and transferrable as possible.

The theme of the second seminar was cooperative and multi-field action on technical systems. The seminar also dealt with the system structures of a technical plant and the action structures relating to them. Aspects of self-training in connection with the problem and system-oriented elaboration of technical contents stood in the foreground of this seminar:

The technical content covered control system tasks for different parts of a transport system. These tasks were to be solved with joint approaches from the fields of pneumatics, switch electronics and PLC/ SPC programming. The project tasks were carried out by mixed teams - teachers and instructors (electrical engineering and metalwork) - as independently as possible. A brief familiarisation of the unknown fields was necessary so as to provide everyone with a common basic understanding of the functional processes and to create a basis for communication. The approaches and experiences were considered from vocational education viewpoints.

The third seminar focused on different aspects and levels of work organisation and the structures in a production company. It involved work on a technical system.

The project tasks included the commissioning of part of a transport system (as a subsystem of an automated production plant) with trouble-shooting and removal. Intertwined problems from the fields of piece-part production (metal-cutting CNC milling and lathing), mechanical assembly, pneumatics, electrical/electronic engineering and PLC/SPC programming was to be carried out within the framework of a fictitious company, partly with real (assembly plant, CNC production) and partly with simulated fields of work. From a methodological approach, it was carried out as an integration of planning game and project work on real objects. The participants monitored their work and learning process and their work organisation in the team independently within the framework of the preset company structures and process steps. The seminar leaders took over the roles of production manager, method engineer, advisers on further training, etc. in the fictitious "company". The individual phases of the project were accompanied by comprehensive rounds of reflection and discussion of the pedagogical, didactic and methodological approaches for system-oriented learning of how to act in a real job situation.

Parallel to the completion of the entire assembly plant for the training and further training of different target groups, company work organisation, the overall technical system and their pedagogical transformation shall be the focus of the advanced training seminars still to be held.

4   Transfer of the Results / Contact Address

The project group publishes a quarterly "PTQ-Info" (distribution: about 15,000 copies) for information and the continuous exchange of experience. It can be ordered free of charge as part of the "BFZ-Info" from the contact address named below.

In the course of the project thus far, specialist conferences were held in 1989 and 1990 to enable the pilot partners to exchange views and experiences among one another and with other firms, vocational training institutes, etc. The PTQ project is permanently presented at relevant specialist conferences and trade fairs. Moreover, PTQ has become a German demonstration project for the European cooperation network EUROTECNET since 1990. Within the scope of EUROTECNET and other programmes, e.g. COMETT, FORCE, etc., a number of contacts have been made which we would like to develop and expand in the future.

Literature:

/1/ Adamowsky, J.; Kluger, J.; Senicar, F.J.: Informationstechnische Grundbildung für gewerblich-technische und naturwissenschaftliche Berufe. Abschlußbericht zum gleichnamigen Forschungsprojekt des BFZ Essen. Final report on the BFZ Essen research project of the same name, Essen 1989

/2/ Bosch, H.: Regionale Entwicklung und Weiterbildung [Regional Development and Further Training]. Cooming out 1992 in: Back, Gnahs (Editor): Weiterbildung als Faktor der regionalen Entwicklung (Information der Akademie für Raumplanung und Landeskunde) [Further Training as a Factor of Regional Development], Hannover 1992

/3/ Corbett, J.M.; Rasmussen, L. B.; Rauner, F.: Crossing the Border. The Social and Engineering Design of Computer Integrated Manufacturing Systems, London, Berlin, Heidelberg et al., 1991

/4/ Feldmann, B.; Kluger, J; Langenbeck, J.: New Starting Points for European Vocational Training. In: Nyhan, B. (Editor): Developing People's Ability to Learn, Brussels, 1991, pp 94-111

/5/ Heidack, C. (Editor): Lernen der Zukunft. Kooperative Selbstqualifikation - die effektivste Form der Aus- und Weiterbildung im Betrieb [Learning of the Future. Cooperative Self Training - The Most Effective Form of Training and Further Training in the Company], Munich, 1989

/6/ Hellwig, H. J.; Richter, H.; Tepper, J.: Arbeitsplatzgestaltung und Qualifikation [Shaping the Work Place and Training]. Workshop Report No. 91 from the "Mensch und Technik - Sozialverträgliche Technikgestaltung" ["People and Technology - Socially Agreeable Shaping of Technology"] program of the state of Northrhine-Westphalia, Düsseldorf, 1991

/7/ Hinte, W.: Non-direktive Pädagogik [Non-directive Paedagogics], Wiesbaden, 1990

/8/ Jungk, R.; Müllert, N.R.: Zukunftswerkstätten [Future Workshops], Hamburg, 1981

/9/ Klaßen, L.; Kluger, J; Meyer, N.; Richter, H.: Qualifizierungsinitiative im Bereich Produktionstechnik [Training Initiative in the Field of Production Technology]. In: Gewerkschaftliche Bildungspolitik, Booklet 10, 1990, pp 233-237

/10/ Meyer, N.; Baron, W.; Feldmann, B.; Kluger, J.: Qualifikation 2000. Thesen zur zukünftigen Qualifikation von Facharbeitern in den elektro- und Metallberufen [Training 2000. Theses on the Future Training of Technicians in the Electrical Engineering and Metalwork Vocations]. In: TH Darmstadt (Publisher): Neue Technologien in der Berufsschule [New Technologies in Vocational Schools], Darmstadt, 1988

/11/ Pätzold, G. (Editor): Lernortkooperation. Impulse für die Zusammenarbeit in der beruflichen Bildung [Learning Site Cooperation. Impulses for Collaboration in Vocational Training], Heidelberg, 1990

/12/ PTQ-Projektinformation [PTQ Project Information] 1.89 - 3.91. The series is continued every quarter (and can be obtained from the BFZ Essen free-of-charge)

/13/ PTQ-Projektgruppe im BFZ Essen (Publisher): Produktionstechnische

88

Qualifikationen im Lernortverbund [Production Technology in Regional Cooperations]. First interim report of the pilot experiment of the same name, Essen 1991

/14/ PTQ-Projektgruppe im BFZ Essen (Publisher): Documentation der technisch-didaktischen Konzeptionen der PTQ-Modellversuchspartner [Documentation of the Technical and didactic Concepts of the Partners in the PTQ Project], Essen, January 1991

/15/ Rauner, F. (Editor): Gestalten - eine neue gesellschaftliche Praxis [Shaping - A New Social Practice], Bonn, 1988

/16/ Rogers, C.R.: Die non-direktive Beratung [Non-directive Counselling], Munich, 1972

/17/ Rogers, C.R.: Entwicklung der Persönlichkeit [Development of the Personality], Stuttgart, 1982

/18/ Tausch, R.; Tausch, A.M.: Erziehungspsychologie [Educational Psychology], Göttingen, 1979

/19/ Volpert, W.: Pädagogische Aspekte der Handlungsregulationstheorie [Pedagogical Aspects of the Action Regulation Theory]. In Passe-Tietjen, H.; Stiehl, H. (Editors): Betriebliches Handlungslernen und die Rolle des Ausbilders [In-company Action Learning and the Role of the Instructor], Wetzlar, 1985

Qualified CIM Training by a Car-Supplier

Heiner Mählck
Panskus Unternehmensberatung
Wuppertal, Germany

Abstract

If you want future, you just have to make it. According to this motto, a new CIM workshop programme was tested and implemented by a subsupplier of the motor industry.

CIM represents an operational aid to put into practice the material and logistics strategy of JIT (Just In Time). This strategy results from the idea to provide information and material at the right place, in the required quantity and necessary quality and at exactly the right point of time.

1   Target: To Understand Connections

During 6 workshop-days 3 groups, of 13 persons each, obtained knowledge on all requirements and measures needed for translating the above mentioned philosophy into action.

Improving product quality, increasing flexibility in fabrication and enhancing efficiency are the most important short-term means for resisting competition and preserving jobs. Demonstrating these relationships was one of the targets of this training initiative.

The suitably trained staff should be prepared at an early date for variations in the company as a result of the implementation of new technologies. Learners, on the other hand, acquire the knowledge, ability and proficiency necessary for active and creative cooperation.

In accordance with the above, the workshop has following contents:

-    Formulation of strategic targets to realise the JIT philosophy.
-    Connections between new technologies and new qualifications.
-    Organisational preconditons and possible practical ways for splitting up production into segments.
-    Central order-related function of a PPS system within the CIM framework.
-    PPS implementation as a complex problem and description of effects and changes in the whole "factory" system.
-    Use of the technical terminology (terms, abbreviations, definitions, etc.) in relation to CIM and JIT.

Flexible, automatic manufacture of Hooke joints, implemented in 1986, represents the technical ideal for the required precision. 6000 Hooke joints are manufactured and tested automatically every day - from the supplied slug up to the precise end product - by a production line working around the clock.

As this modern technology spreads, the staff should participate actively in the organisation process. In this case different problems, not only those of a technical nature, will have to be solved. The realisation of innovative CIM projects requires the mastering of new technologies as well as a comprehensive learning process, extensive enforcement activities, persuasive power, the settlement of conflicts and reduction of acceptance problems.

Thus, supplementary behavioural training, in addition to the imparting of special knowledge, is of immense importance. CIM training is efficient for entire whole staff concerned with computer integration. All participants must have special knowledge which includes the philosophy, strategy and tools of computer integration. Moreover, they have to be able to cooperate in a practical and social manner. One of the focal points in this connection was the sensitiveness related to overlapping thinking and managing.
The most important qualifications for working in a 'company with a future' are the ability to think beyond one's own field, basic knowledge about contiguous branches and the ability to see problems which are not part of the area in question and solve them through cooperative efforts.

A workshop prepared in such a manner offers the best preconditions to qualify the staff for integrated capabilities as well as to deal with new technologies (see Fig.).

It was also helpful that the participants came from different departments and from various levels of the hierarchy, i.e. middle and lower management. Thus everyone involved obtained detailed insight. Besides numerous interdisciplinary discussions with regard to technique and organisation, the integrated capabilities were improved during the practice-oriented teamwork.

The workshop was divided into 2 sections of 3 days each. The first part mainly dealt with basic knowledge while the second involved intensive work with CIM modules.

Subjects in detail:

First part:

- Factory and Office in the Process of Change
- Training and Characteristics in the Process of Change
- Teamwork / Treatment of Problems / Diagrams and Graphs
- World of Facts for Electronic Data Processing
- Offer / Order

Second part:

- CAD - Computer Aided Design
- CAP - Computer Aided Planning
- CAM - Computer Aided Manufacturing

Order-related component

- PPS - Production Planning System

An organised course took place every day. The morning was reserved for presentations supported by graphs and videos. After a short test in the afternoon concrete problems had to be solved in teamwork.

Result : Identification with the Company's Objective

The acquired theoretical knowledge was employed in practical situations. This workshop achieved the following results:

- Participants understood very quickly and were able to put their theoretical JIT and CIM knowledge to practical use. This can be attributed to the intensive inclusion of all participants during the whole course.

- The identification with the company's objectives and philosophy could be intensified by practice-oriented teamwork.

- The close contact within the group brought about greater appreciation of difficulties and problems that exist in other parts of the company.

- 'Together we are strong'. Through the cooperation of the staff this makes it possible to realise bigger projects in the course of organisational changes.

- With regard to technical, social and methodological aspects the competence of all those involved was enhanced.

- By reporting the results of teamwork, the participant's self-confidence could be strengthened.

This kind of workshop organisation succeeded in showing the importance of domain overlapping thinking and management as a basis for the technical reorganisation of production sections.

This facilitates the realisation of a transparent transmission of material and information as one of the most important components in the course of JIT/CIM-oriented production. Thus a far-sighted and stable foundation was created for "... making a company's future".

CIM Training Concept for Small and Medium-Size Enterprises (SME)

Markus Nüttgens; Prof Dr. August Wilhelm Scheer
Institut for Economic Informatic, University Saarland
Saarbrücken Germany

1   Training - a "CIM Bottleneck"

The application of computer-aided technologies in production and the related idea of Computer Integrated Manufacturing (CIM) will decisively influence the competitive capacity of small and medium-size enterprises (SME). CIM means computer-supported processing of integrated operational processes between production planning and control, design, process planning, operations planning and scheduling, production and quality assurance. In this way CIM is a strategy which aims at the information-related linking of various operational areas of an enterprise [1].

Production problems can be reduced to the following denominator: according to the market, enterprises must produce innovative and high quality products while reducing the "time to market". Besides these market demands the enterprise is internally confronted with increasing costs, time pressure and higher quality standards. Both time and quality as well as productivity and price influence the competitive capacity. Short and reliable delivery dates, flexible and short-term reaction on customer wishes, and guarantee of high quality are required and this pressure will intensify even more.

Enterprises having up to 500 employees and annual net sales of up to 25 million marks belong to the SME group [3]. Not only are they found different sectors, they also have a strong heterogeneity regarding the product range, the type of production, organisation and customer-supplier relationship. Especially the close customer-supplier relationship increasingly requires the application of CIM components. For many SME's this can become a question of survival [4].

The firm's objectives connected with the use of integrated information systems cannot only be regarded as aspects of information technologies. Moreover there are interdependent technical, organisational and personnel aspects. Managers as well as decision makers of the SME's often misunderstand that the "CIM capability" of an enterprise must take the decisive hurdle here and that personnel development planning is not the last link but is becoming more and more a "bottleneck factor" [5]. Therefore, a model for integrated CIM training must fulfill the requirements of target-group-oriented personnel development. It should be used for a systematic derivation of training courses in a CIM environment.

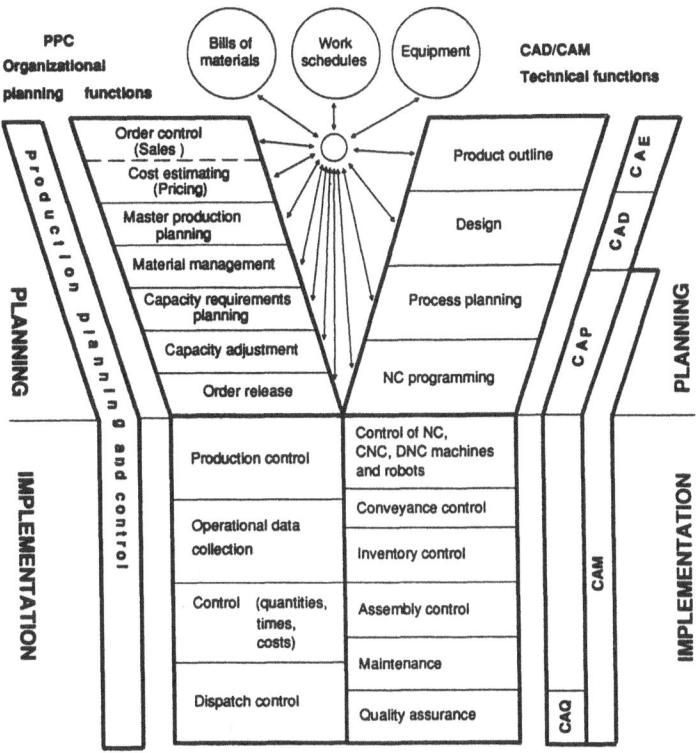

Fig. 1 Information systems in production

2 Draft of a CIM Training Model

CIM training - in the sense of a long-range initial and further training - cannot only be based on short-term modifications of existing initial and further training programmes. Instead strategies and models must be worked up from which modern training programmes can be derived [6].

2.1. Design Strategy and Models

An overlapping CIM training concept can be deduced using either [7]:

- a top-down strategy which hierarchically deduces training contents from the view of an enterprise-wide system and assigns them to the target groups or
- a bottom-up strategy which analyzes the specification profiles in the

various sub-areas and effects an enlargement within the same or even for higher hierarchical levels.

To achieve an overlapping training concept, it is necessary to use the top-down strategy in the first stage of the structuring. Therefore, a model is developed which aims to cover both personnel tasks and personnel structure. This model files or localises existing or lacking training programmes. Based on these programmes, more and more detailed training subjects will be derived in the further stages of the structuring and be documented in requirement profiles. These subjects can be used to show the changes in task structure and subjects. In contrast to the top-down analysis, the design of a model is not applied in the usual bottom-up analysis. For this reason, it can only be used for the modification of already existing training courses. Current training offers in a CIM environment may be explained by the usual bottom-up-oriented strategy. But they have two essential disadvantages:

-   Because of an inadequate target group differentiation, CIM training courses do not exist or they are insufficient in didactics, methodology and substance for managers.
-   Since there is no embedding in a training model, interdependencies between the individual training courses cannot be taken into account. General user training first of all imparts more special knowledge with regard to isolated tasks. Up to now only rudimentary approaches covering all areas exist.

For this reason, the CIM approach with its integrative idea requires the design of a model from which concrete training courses can be derived. In the following part a CIM training model is described which especially considers the training needs of the SMEs.

2.2. CIM Training Model

Far reaching changes caused by the use of integrated information technologies also require altered qualifications. On the one hand, for those whose task subjects will be directly changed because of new equipment and new defined task rules; on the other hand, for those who decide on and plan the use of the technology [8]. Using the target group differentiation and a step-by-step-strategy, a CIM training model can be derived. Its training courses aim to achieve the following basic levels of CIM training:

-   Application knowledge: required training to handle CIM components and use them efficiently.
-   Decision-making knowledge: required training for enterprise-wide CIM planning and realisation.

The basic qualifications for CIM, described above, can be shown in an information pyramid. In designing a CIM training model, the intention was to develop an open-structured skeleton concept. This means concrete training programmes (basic curricula, training careers, didactic equipment, etc.) and its organisational execution (initial and further training, training on the job, etc.) depend on the target group and on the availability of training potential.

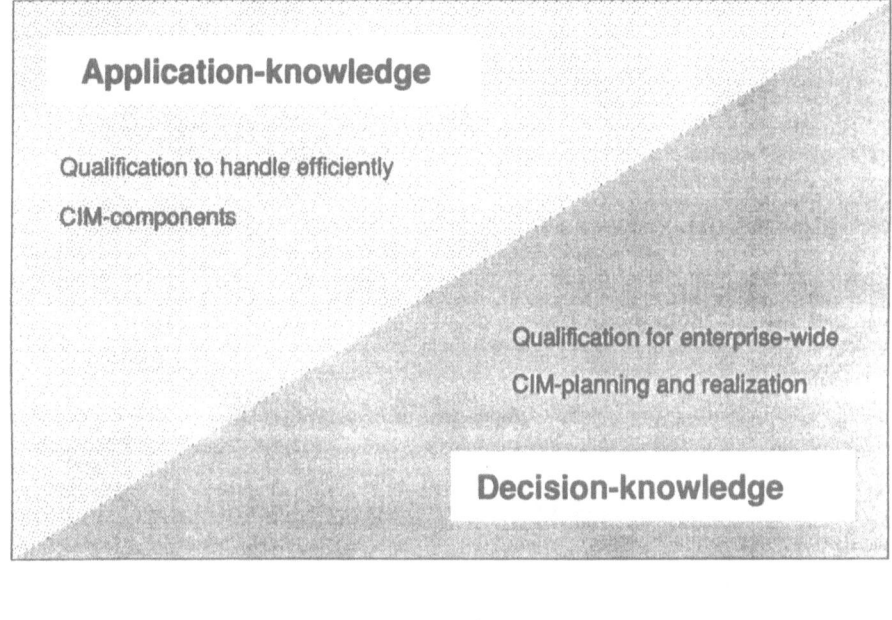

operator   ·········· ···········   *top management*

Fig. 2   Basic levels of CIM training

Furthermore, on the basis of this training model, concrete training courses can be derived and used in the course of adequate personnel development planning. This model provides systematic CIM training. In analogy to the "technical" integration of EDP systems and the "organisational" integration of tasks, a "training" integration is to be achieved.

At the level of component training the distinction between general user training and specific user training is very important. General user training should impart product-independent knowledge as far as possible. This refers to training which will be carried out with the aid of exemplary hardware and software configurations. Specific user training carried out by manufacturers and suppliers should not and cannot be replaced by this general user training. But they have particular importance because of a one-sided orientation of specific user training towards special hardware and software components.

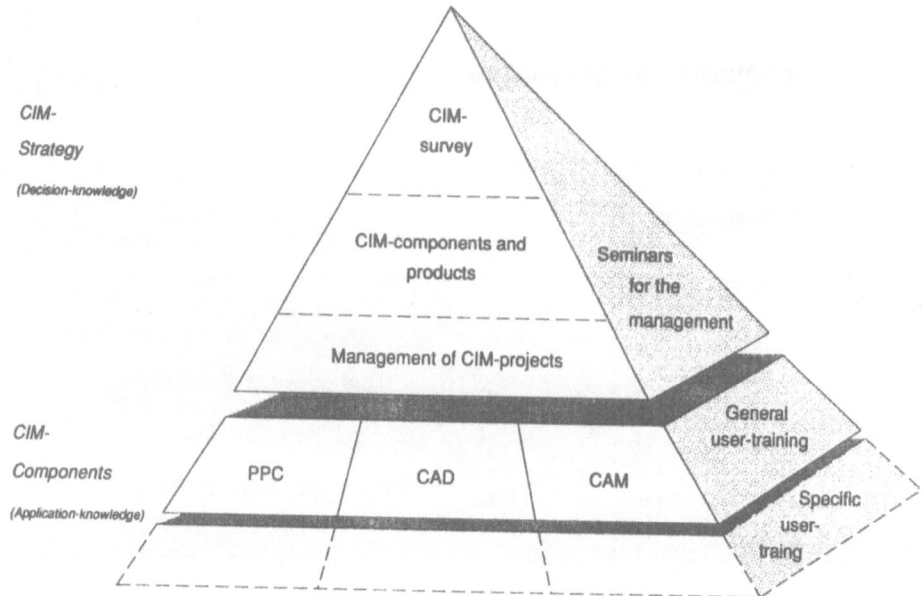

Fig. 3   CIM training modell

3.   Development of Training Courses

Based on the training model, specific training courses can be derived in the form of courses, seminars, etc..

When compiling new training programmes, the first step is a detailed analysis and a stipulation of the training concept. A training concept should include training aims, subjects and methods. Based on this training concept training materials can be provided which are more or less constant. Nevertheless, they must be updated constantly.

3.1   Training Topic "CIM Strategy"

The "CIM strategy" training programme is especially created for managers who decide on implementation of EDP systems in their enterprises. These managers increasingly have to estimate investments related to computer-supported manufacturing though they are already overtaxed by the constantly and rapidly evolving process of innovation. This process will intensify dramatically. The continuous decrease of the "half-life of knowledge" causes a shift from initial training to further training. Long-life learning for all hierarchical levels of the enterprise is the answer to the challenges which can no longer be met with initial education and training. [10].
The aim of the training is to provide seminar participants with decision-making

CIM - Computer Integrated Manufacturing

- Moduls of a sequence of seminars "CIM-Strategy for SME" -

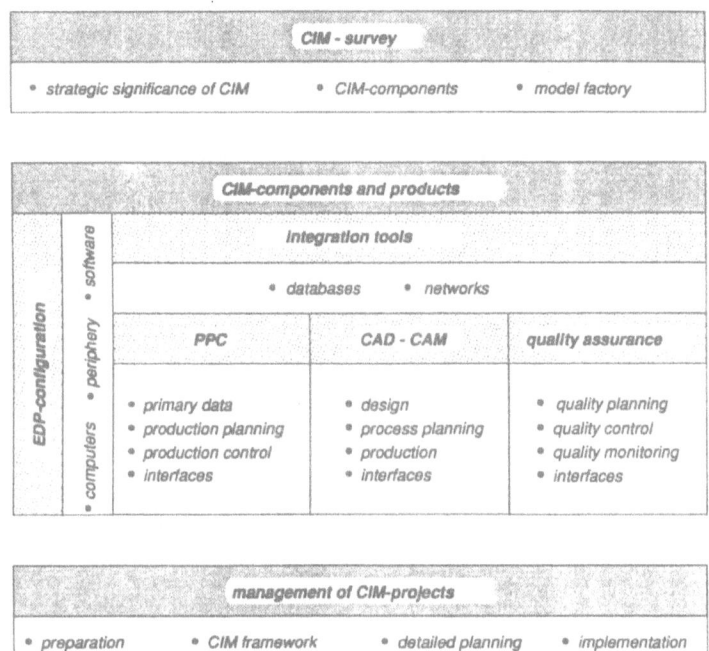

Fig. 4   Subjects of seminars "CIM strategy"

support and basic strategic knowledge of the implementation possibilities of CIM components and their integration in an operational enterprise-wide concept.

Comprehensive training with the emphasis on "CIM strategy" especially designed for managers might be subdivided into three modules and include the following topics:

Module 1:   CIM Survey

Here the participants become familiar with the terms, concepts and effects of the CIM idea. The fundamentals of computer-supported production are taught and the strategic meaning of CIM planning is pointed out.

Module 2:   CIM Components and Products

Here the single CIM components their interdependencies and their significance are explained. Decision-making support which can be used for the selection and the use of component-specific hardware and software is provided. The module "CIM components and products" is subdivided into the following seminars:

- Configuration of EDP systems
- Integration tools
- Production planning and control
- CAD and CAD-CAM
- Quality assurance

Both the fundamentals and the products will be dealt with in each subject.

Module 3:   Management of CIM projects

Here, a planning model is illustrated which serves as orientation support in the systematic implementation of CIM components. Taking into consideration organisational, personnel and juridical aspects, typical problems are illustrated and analysed.

Parallel courses

To complete single seminars, parallel courses are recommended. They should be a bridge between "theory and practice". Moreover, a lively exchange of experience should stimulate discussion centering around a concrete example.

Typical subjects of these courses could be:
- Study trips to technology transfer centres and visits to model factories.
- Study trips to enterprises which already use CIM technologies.
- Experience reports of enterprises.
- Reports of management consultants as well as demonstrations shown by suppliers of CIM components.

The concrete planning and realisation of such parallel courses depend on the respectively available resources.

Besides the learning objectives and subjects, the training concept includes the learning methods. By developing manager training, the following basic conditions must be taken into account [11]:

Time schedule:
- Compact knowledge procurement in evening or weekend seminars.
- Broad time frame to make knowledge and experience transfer possible.
- Different study trips and CIM examples (reference enterprises, model factories, etc.).

Material:
- Supply of CIM concepts that are not enterprise- or manufacturer-specific.
- Distribution of high-quality training materials, i.e. qualified compendium for one's own use (transparencies).
- Essential equipment includes an overhead projector, video and flipchart.

Instructors:
- Competent instructors from theory and practice in a balanced proportion.
- The complex subject requires experts for each topic.

Workshop character:
- Small groups (approx. 15-20 participants).
- Ensuring a discussion forum between instructor and participant.
- Appealing surroundings.

3.2 Training Topic "CIM Components"

This training programme is especially designed for the user of CIM components. The user's task is the treatment of sub-processes within order handling. Here, the EDP-supported tools can be related to the more technical (CAX systems) and operational (PPC systems) task areas. The use of CIM components gives rise to job enrichment and job enlargement at the workplace:

- Job enrichment include the control of the implemented technology and the required operational techniques (functional qualifications).
- Job enlargement is the result of the change of operational processes and a higher competence (extra-functional qualifications).

The required training can therefore be analyzed functionally or extra-functionally.
Based on the CIM training model, training programmes can be differentiated regarding CIM components in general and specific user training.
In training courses that are not product-specific basic knowledge of the CIM components' functionality in a specific area of applications is imparted. One focus of these training courses is on supra-operational knowledge. General user training might consist of three course modules:

Module 1:   Fundamentals of EDP technology

Here EDP-related fundamentals of how to use computer supported equipment are imparted. The participant learns how to use the hardware components and the software of the system. Since these fundamentals are largely independent of the area of application and the functional range of its specific CIM component, the module "EDP fundamentals" is formulated independent of application.

Module 2:   CIM Component Training

Here the emphasis is on the procurement of knowledge of computer-aided equipment in specific applications. Solid knowledge of the functional range of the CIM components and the underlying working procedures is imparted. To co-ordinate the user courses, the CIM training model distinguishes between the following areas of application:

- Production planning and control (PPC)
- Computer-aided design (CAD)
- Computer-aided manufacturing (CAM)

In this model quality assurance is assigned to the CAM area. Since more and more importance is attached to quality assurance, it can also be integrated into the model as an independent application area.

Module 3:   CIM Interfaces

Here the relationship between CIM components is discussed. The integration of CIM components within an operational enterprise-wide system is shown to the participants. Afterwards they will be able to recognize the technical and operational functions of order handling. Using CIM sub-chains, the essential interfaces between the CIM components are shown: [12]:

-     Subchain 1:   Linking production planning and control.
-     Subchain 2:   Linking CAD and CAM
-     Subchain 3:   Linking primary data management
-     Subchain 4:   Linking operational data collection and CAM
-     Subchain 5:   Linking inter-company data exchange

Users of CIM components are usually not university graduates and have had practice-oriented training. Special examples of computer-supported order handling help in teaching these training subjects. The use of concrete working tasks provides for deliberately controlled working and learning action (action-oriented learning). These working tasks should support the course participant in stabilising his working action by constructing cognitive mental models [13].
Besides exercises on concrete CIM components the use of new media (simulation, animation, hypermedia) is becoming more and more important for didactic preparation of training programmes.

4   Supporter of CIM Training

The main supporters of the external CIM training programmes are:

-     universities and academies,
-     chambers and associations,
-     commercial enterprises.

It is useful to retain the training programmes of these institutions because

-     training within a company might be not profitable since there are only a few participants,
-     external institutions have more comprehensive competences and can offer appealing basic conditions,
-     external training programmes permit those involved to meet and exchange experience,
-     training within a company is linked to the available equipment,
-     external training permits a consistent separation of training and daily work,
-     special courses, including officially recognized graduation, cannot be supplied by the company,
-     innovative technologies also need a fresh impetus from the "outside".

Seminars for Decision Makers and Managers

The construction of CIM technology transfer centres in Germany within the scope of the BMFT Programme "Production Technology 1988-1992" is an important contribution to the extension of the future "widely effective CIM Technology Transfers" between universities and SME's. About 20 CIM demonstration sites have been constructed at various universities to give examples for CIM solutions, to offer training programmes, and to provide guidance independent of manufacturing sector [14].
Technology transfer is especially designed for decision-makers and managers of the SME's. Cooperation and guidance among chambers and associations to create both enterprise-wide training concepts and training for instructors are increasingly gaining in significance.

General User Course "CIM Components"

Organisations like the chambers of handicrafts, the Chambers of Industry and Commerce, trade union organisations, etc. offer user courses for technical fields (CAX components). They make efforts to integrate CIM sub-chains (e.g. the relationship between CAD and CAM) within their training concepts. Training programmes with production planning and control approaches already exist but they usually do not contain a general enterprise-wide concept of a PPC system.
To develop a satisfactory training programme, it is necessary to incorporate the course modules into an integrated training concept which then allows step-by-step adaptation and extension.

5 Comprehension and Prospects

The training model shown should not be regarded as an approach which deals with still remote subjects. Especially in rapidly changing times instructors for initial and further training are asked to assess their training requirements carefully. Only those who have reliable information can systematically plan and economically manage such a modern training programme.

The described training concept for SME's has been developed by the Institut für Wirtschaftsinformatik (IWi), Universität des Saarlandes, within the scope of COMETT I, a research project supported by the EC. At present this concept is further developed in cooperation with the chambers of craft trades of Luxembourg, Saarland and Trier. Based on this concept, a training model within the "CIM strategy" for managers was developed in 1990 and is currently being tested by the project partners. Moreover, in COMETT II, a training model for production planning and control (Project 3017/Cb "PPC in small and medium enterprises") is being developed for, among other things, the application of a hypermedia-based tutorial.

Experience has shown that the CIM training model ensures a target-group-oriented and consistent derivation of CIM training courses. To get a systematic CIM training this model is also an aid for the integration of training programmes analogous to the "technical" integration of EDP systems and the "organisational" integration of tasks. Enterprise-wide training concepts ensure that CIM users and CIM managers can be trained according to their specific standards. Technological shortcomings can be compensated for by means of capital, shortcomings in training cannot. Many participants have this bitter experience.

Appendix

[1]: A more detailed description see Scheer, A.-W. (Bd.-Hrsg.): CIM-Strategie als Teil der Unternehmensstrategie, in der Reihe: Bey, I. (Hrsg.): CIM-Fachmann, Köln 1990.

[2]: Scheer, A.-W.: CIM (Computer Integrated Manufacturing) - Der computergesteuerte Industriebetrieb, 4. Aufl., Berlin-Heidelberg-New York u. a. 1990, p. 2.

[3]: See Kommission der Europäischen Gemeinschaften, Generaldirektion XII (Hrsg.): Forschungs- und Technologieförderung der EG, o. O. 1989, p. 13.

[4]: See Scheer, A.-W.: CIM - eine Herausforderung für den Mittelstand, in: Scheer, A.-W. (Hrsg.): Computer Integrated Manufacturing - Einsatz in der mittelständischen Wirtschaft, Berlin-Heidelberg 1988, p. 3ff.

[5]: See Scheer, A.-W.: Der Mittelstand - der ideale CIM-Anwender?, in: Scheer, A.-W. (Hrsg.): CIM im Mittelstand, Berlin-Heidelberg-New York u. a. 1989, p. 2ff.

[6]: See Nüttgens, M.; Eichacker, St.; Scheer, A.-W.: CIM-Qualifizierungskonzept für Klein- und Mittelunternehmen (KMU), Saarbrücken 1991 (Veröffentlichungen des Instituts für Wirtschaftsinformatik, No.75, Herausgeber: A.-W. Scheer).

[7]: See Deutsches Institut für Normung e.V. (Hrsg.): Normung von Schnittstellen für die rechnerintegrierte Produktion (CIM) - Standortbestimmung und Handlungsbedarf, Berlin-Köln 1987, p.182.

[8]: See Karl, P.: Aus- und Weiterbildung - Voraussetzungen für eine erfolgreiche CIM-Implementierung, in: Scheer, A.-W. (Hrsg.): CIM im Mittelstand, Berlin-Heidelberg-New York u. a. 1990, p. 97-113.

[9]: Nüttgens, M.: Qualifizierungskonzept für CIM in Klein- und Mittelunternehmen, in: Einführungsstrategien zur CIM-Impelementierung in Klein- und Mittelunternehmen, Proceedings zur Fachtagung am 11. Mai 1990 in Luxemburg, o. O. u. J., p. 89.

[10]: See: Scheer, A.-W.; Keller, G.; Nüttgens, M.: Integrationsschwerpunkt "CIM-Qualifikation" - Entscheidungswissen ist gefragt, in: Personal, 42(1990), No.6, p. 244-249.

[11]: See Hahn, R.: CIM-Training für Führungskräfte - Aufgaben - Lösungen - Erfahrungen, in: Bullinger, H.-J. (Hrsg.): Produktionsforum'88 - Die CIM-fähige Fabrik, Berlin-Heidelberg-New York u. a. 1988, p. 605-617.

[12]: See Scheer, A.-W.: CIM (Computer Integrated Manufacturing) - Der

computergesteuerte Industriebetrieb, 4. Aufl., Berlin-Heidelberg-New York u. a. 1990, p. 57ff.

[13]: See Thomfohrde, A.: Integration und Qualifikation im Funktionsbereich Konstruktion, in: Bullinger, H.-J. (Hrsg.): CIM-Integration und Qualifikation - Berufliche Bildung im Technologietransfer, Köln 1989, p. 109ff.

[14]: Karl, P.; Geib, T.: Das Programm der CIM-Technologie-Transfer-Zentren am Beispiel des Standortes Saarbrücken, in: Scheer, A.-W.: CIM (Computer Integrated Manufacturing) - Der computergesteuerte Industriebetrieb, 4. Aufl., Berlin-Heidelberg-New York u. a. 1990, p. 252-259.

Literature

/1/ DIN (Hrsg.): Normung von Schnittstellen für die rechnerintegrierte Produktion (CIM) - Standortbestimmung und Handlungsbedarf, Berlin-Köln 1987.

/2/ Hahn, R.: CIM-Training für Führungskräfte - Aufgaben - Lösungen - Erfahrungen, in: Bullinger, H.-J. (Hrsg.): Produktionsforum'88 - Die CIM-fähige Fabrik, Berlin-Heidelberg-New York u. a. 1988, S. 605-617.

/3/ Karl, P.: Aus- und Weiterbildung - Voraussetzungen für eine erfolgreiche CIM-Implementierung, in: Scheer, A.-W. (Hrsg.): CIM im Mittelstand, Berlin-Heidelberg 1990, S. 97-113.

/4/ Karl, P.; Geib, T.: Das Programm der CIM-Technologie-Transfer-Zentren am Beispiel des Standortes Saarbrücken, in: Scheer, A.-W.: CIM (Computer Integrated Manufacturing) - Der computergesteuerte Industriebetrieb, 4. Aufl., Berlin-Heidelberg-New York u. a. 1990, S. 252-259.

/5/ Kommission der Europäischen Gemeinschaften, Generaldirektion XII (Hrsg.): Forschungs- und Technologieförderung der EG, o. O. 1989.

/6/ Nüttgens, M.: Qualifizierungskonzept für CIM in Klein- und Mittelunternehmen, in: Einführungsstrategien zur CIM-Impelementierung in Klein- und Mittelunternehmen, Proceedings zur Fachtagung am 11. Mai 1990 in Luxemburg, o. O. u. J., S. 77-95.

/7/ Nüttgens, M.; Eichacker, St.; Scheer, A.-W.: CIM-Qualifizierungskonzept für Klein- und Mittelunternehmen (KMU), Saarbrüken 1991 (Veröffentlichungen des Instituts für Wirtschaftsinformatik, Heft 75, Herausgeber: A.-W. Scheer).

/8/ Scheer, A.-W.: CIM - eine Herausforderung für den Mittelstand, in: Scheer, A.-W. (Hrsg.): Computer Integrated Manufacturing - Einsatz in der mittelständischen Wirtschaft, Berlin-Heidelberg 1988, S. 1-16.

/9/ Scheer, A.-W.: Der Mittelstand - der ideale CIM-Anwender?, in: Scheer, A.-W. (Hrsg.): CIM im Mittelstand, Berlin-Heidelberg-New York u. a. 1989, S. 1-15.

/10/ Scheer, A.-W.: CIM (Computer Integrated Manufacturing) - Der computergesteuerte Industriebetrieb, 4. Aufl., Berlin-Heidelberg-New York u. a. 1990.

/11/ Scheer, A.-W. (Bd.-Hrsg.): CIM-Strategie als Teil der Unternehmensstrategie, in der Reihe: Bey, I. (Hrsg.): CIM-Fachmann, Köln 1990.

/12/ Scheer, A.-W.; Keller, G.; Nüttgens, M.: Integrationsschwerpunkt "CIM-Qualifikation" - Entscheidungswissen ist gefragt, in: Personal, 42(1990), Heft 6, S. 244-249.

/13/ Thomfohrde, A.: Integration und Qualifikation im Funktionsbereich Konstruktion, in: Bullinger, H.-J. (Hrsg.): CIM-Integration und Qualifikation - Berufliche Bildung im Technologietransfer, Köln 1989, S. 95-116.

The Teaching of Key Qualifications Using Integrated Learning Stations as Shown in the Example of the Flexible Learning Laboratory System (FLS)

Prof. Dr. -Ing. Walter E. Theuerkauf; Andreas Weiner
IAET, University Hannover
Hannover Germany

1   Introduction

The successive introduction of new technologies into industry and administration has led to an altered form of operational rationalisation, which has now, after the technical phase of automation, resulted in altered work structures. These show that, in spite of technical problem-solving, a "New Way of Thinking" is beginning to gain acceptance. The concepts connected with it are characterised by the abandonment of the Taylor-Ford dogma and a turn towards an integral use of the capacity for work, i.e. away from mere performance and towards object-oriented job content./1/

Altered training requirements in the company environment, which as a rule are connected with the introduction of computer-controlled procedures and equipment, have led to consequences for initial vocational training and company/industry-wide training. As a result of this we find learning processes whose content and methods have been changed and which put the training emphasis not on the development of technical competence, as hitherto, but extend it with regard to methodology and social competence.

Key qualifications are regarded as an important maxim of current concepts of personnel economy and education strategy. Systemic acting and thinking is an important key qualification which it is necessary to develop during vocational training./2/

The altered training requirements in industry have been taken into account in the reorganisation of the metalwork occupations. New job outlines with new technical content and methodological requirements have been created which reflect the changes in technological and work organisation requirements in industrial manufacturing./3/

2 The significance of key qualifications and key content for the Industrial Mechanic

Studies are available on the Industrial Mechanic/ Production Mechanic regarding his function of machine supervision and maintenance in industrial scale manufacturing./4/ These studies have been examined with regard to training for flexible automated manufacturing equipment. The workplaces analyzed are considered to be exemplary for cutting and non-cutting manufacturing. Manufacturing systems with varying levels of automation were hereby taken into account. The training requirements were determined for both normal operation of the machinery as well as for faults during the production process and for technical faults. Competence in the handling of flexible automated production machinery is regarded here as an example of systemic acting competence./5/

In view of the job characteristics with respect technical organisation and personnel management, the spectrum of tasks for a machine supervisor encompasses, along with the typical operator and supervisor tasks, setting up and retooling work as well as the correction and/or optimisation of process programs connected with test runs, preventive maintenance work and the localisation and isolation of errors.

Observation and supervision of situations and processes as well as signal recognition number among the cognitive requirements for machine supervisors of complex manufacturing systems during the diagnosis and correction of errors. In error situations special demand are made on the observation of machine states and product features in order to localise the error. The occurrence of a fault during the functional operation of the machine is indicated by means of optical and acoustic signals and the corresponding processing situation is localised. By means of further observation of optical signals on the corresponding station and the knowledge concerning the relation between the signals and certain switching states, it is possible to perform local correction of the fault, such as faults in the supply of materials or robot control mechanism.

Within the framework of fault diagnosis and management in complex automated production machinery the skill and knowledge requirements of control technology play a special role. The employee must have a basic knowledge of the interaction between electronic control apparatus and working/switching elements in order to be able to perform fault diagnosis. Knowledge of the operator and switching elements are of special importance. The control technological functional process, represented by means of a lighted display, must be related to the corresponding processing steps. Actual knowledge of programming in stored program control systems is not of significance for machine supervisors during the performance of their job. However, knowledge of program structure, codes/symbols of a programming language should be present. The knowledge of CNC technology refers to the operation and handling as well as the optimisation and adaptation of programs. Programming is not necessary, however. In the case of machining operations, knowledge is required of the crucial parameters and processing peculiarities of turning, milling and drilling processes. Knowledge of industrial robots is restricted to the principle structure as well as handling and operating.

Apart from the training analysis already described the training regulations can assist the orientation of the content structure of teaching in technical colleges.

The range of these activities also requires the acquisition of methodological and social competence in addition to the key content described here. This competence is subsumed, particularly in literature, under the concept of key qualifications.

REETZ describes the key qualifications taking into account ROTH's /6/ personality model. He differentiates between such "behaviour preparedness in the form of capabilities, attitudes, opinions" which biographically come before professional activity and those which affect the professional activity itself. With reference to the latter which are relevant for the interrelation discussed here, REETZ distinguishes between :

-       personal fundamental abilities of character (attitudes, normative orientation, character traits, such as stamina activity, initiative, eagerness to learn),
-       performance, activity and task-oriented abilities (e.g. problem solving, decision-making, concept development),
-       socially-related abilities (the ability to cooperate with others, overcome conflicts and negotiate etc).

Here the question arises concerning the content of vocational training with which it is possible to attain key qualifications. On this topic REETZ states :

"In the field of industrial technology specialised knowledge and skills have not lost any of their significance. At the moment, however, a restructuring is taking place within the scope of the specialised knowledge due to the influence of new technologies, whereby the significance of the amount of planning and action-related knowledge is increasing within the specialised knowledge."

The question still arises concerning the methodological structure of the learning processes with which it is possible to attain key qualifications.

Learning arrangements are required here which link the situative complexity to the action-related orientation, whilst at the same time furthering motivation and encouraging independent learning. Problem-solving is a variant of action to which special importance is attached. Discovery learning plays a special role here : problem-solving is carried out by means of experiencing the discrepancies between an existing scheme and one which has to be learned. The problem is solved via the next higher level of complexity. In this way it is possible to attain an increase in the complexity of thinking and acting which corresponds exactly to the demands of the key qualification problem.

3   Flexible Learning Laboratory System as an instrument for planning

With regard to the media, the given objectives, which are determined by the relatively abstract key qualifications and defined by the learning strategies already known and yet to be extended, also require conformity with the firms' requirements, on the one hand, and the vocational training content defined by the regulations, on the other hand. As a result of this, an open media configuration of a learning system is required which exhibits flexibility in the sense stated here. If the planning of learning processes is based on key qualifications, especially with the argument that they are the best way of coping with changing qualifications due to technological progress, then this consideration will have been translated into a specification for the media. The Flexible Learning Laboratory System (FLS) has worked up this concept and its media technological intention has always represented a planning system which has been designed mainly for integrative learning processes.

The basic qualifications in the individual sectors of New Technologies and how they have arisen due to the subject matter which has to be learned, such as pneumatics, CNC technics and stored program control, suggest themselves as the starting point for a planning system. The media available within a considerable range of variations on the market for educational and learning aids make it possible to solve, on an isolated basis, complex task which have to be performed in a computer-integrated factory. These media teach basic knowledge of basic functions, such as transportation, processing, handling, storage, testing and measuring.

If these individual operations are linked up into an integrated manufacturing unit, it is possible to interconnect, for example, the spatial movement of materials, feed and clamping motions, and the storage of materials in such a way that a complex media system is created which can be composed of both operating equipment of a model character as well as components of an industrial character. The fundamental objective of the planning system, therefore, is to combine the existing media with the multi-media system with variable system complexity created for the respective learning task. Variable means the complexity can be predetermined by vocational and company training, on the one hand, and by the existing media, on the other hand.

The Flexible Learning Laboratory System has, as it were, given itself the pedagogical and didactic objective of offering instructors and students the possibility to generate genetically a system-oriented complex media situation. This can be developed using individual elements as well as existing learning stations. Emphasis on structural relevant learning and the associated thought processes in chains of operations develops a creative learning process on the object involved which does not simply start out from a predefined media configuration of automation oriented to the means of production, rather it allows this configuration to be created. A learning process or media situation which activates both the instructor and the student, thereby permitting individual, possibly more efficient teaching strategies, for example, through a more optimal adaptation to the target group, represents initiative and thus provides a way out of the rut of training with a course-like character. At the same time the possibilities for realising the project's aim are taken into account in an adequate manner, namely, even when more complex, interlocking operations arise through the interconnection of learning stations.

While the previous observations concentrated on the aspect of automation oriented to the means of production, one must also take into account the product-oriented point of view./7/ Production orientation means that the manufacturing process can be depicted in its fundamental phases according to a holistic point of view. In this way the object to be constructed, i.e. the product, is the starting point for the media configuration to be drawn up, i.e. the manufacturing setup. The drawing up of the media configuration itself now assumes the character of a project, which results in the design and construction of media for solving the overall task. Automation thereby represents a subject-oriented work content and a changed dimension of work and technology. It can therefore reflect the foreseeable change from the present technocentric to future anthropocentric industrial work./1/

However, in order to be able to implement such a pedagogical/didactic basic idea, it is necessary to create the prerequisites for the desired interconnection of the learning stations. These are necessary for both the hardware and the software. At the same time, the central aim during the design of the FLS was to create a control system package for various media configurations which is as open as possible and which allows a multifaceted, structural and partial representation of a computer-integrated manufacturing system. The task of this control system is to coordinate operations and, by means of the generation of differing work organisations, enable one to become familiar with their logistical principles. Only when this requirement is fulfilled can the information concerning the simulated manufacturing process be treated with regard to its content in an accurate and decision-oriented manner for the development process to be optimised. In doing so it is possible to effect transfer of planning action models to the production factor information, which, according to management, has the same status as the factors of human capital, operating equipment and materials.

If one orients oneself to the various structures/levels of the decision-making pyramid and to the data flow of a company, this means that the adaptation of the actuator and sensor systems to the stored program control system must be carried out at the lowest level of the work process chain. This different working equipment must be interconnected into a DNC network at the next higher level. If product orientation is assumed here, then it is necessary to utilise all the creative possibilities within the work organisation. It must be possible to realise both assembly-line and workshop production in order to teach the spectrum of work content in such a way that it is possible to clearly accentuate the relationship between work and technology. In order to satisfy these didactic/pedagogical demands, the FLS planning system must be designed in such an open manner that it is easily possible to perform hardware and software alterations at the interfaces to the operating equipment.

4 Controlling technical interconnection of the learning stations

In order to control the processing and material-flow activities, a Flexible Learning Laboratory System requires a concept which enables integration of the various operating equipment in a relatively simple manner. Control means the exchange of data. Difficulties always arise whenever the protocols from different manufacturers are incompatible. This becomes clear when one examines the sample configuration developed by the institute in which the manufacturing cell "drilling" contains a FESTO control unit (FST 404) which is not compatible with the robot made by

MITSUBISHI.

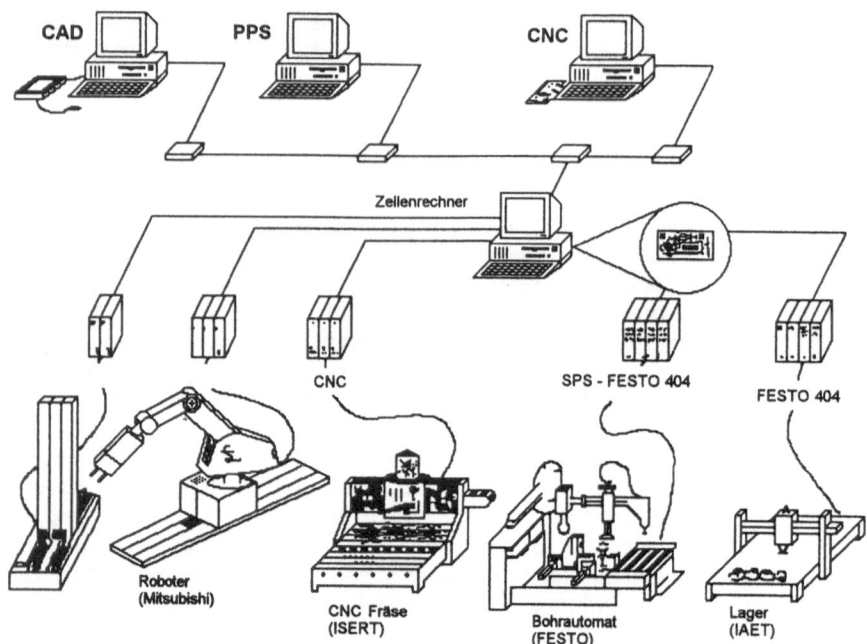

Fig. 1 FLS configuration

There are a series of interconnection variations which are suitable for solving this problem, including LAN which is now offered by every manufacturer of automation equipment, especially for the SPC. These existing solutions correspond, on the one hand, to the present industrial standard while, on the other hand, they require a communications processor for each of the analog learning stations, thus guaranteeing communication within the network (master-slave), including the corresponding network software. In the long term, a solution similar to the SINEC H1 network should be striven for, because the cost situation and bus development will make this solution necessary in view of the emerging standardisation in the field of stored program control. A network concentrator which allows the learning stations to be interconnected in a star configuration has been selected as an alternative and reasonably-priced solution for training purposes, which corresponds to the given objectives and is appropriate for the target group. This concentrator, which can also be considered as part of a network concept, facilitates communication using the specific operating system data protocols of the learning stations. As we assume that the individual operating equipment possesses standard interfaces such as RS-232 or Centronics,

it is possible to set up communication using a PC via its available or extendable number of interfaces. The extension can be carried out using two I/O boards that permit interconnection of a total of eight different pieces of operating equipment. A further advantage can be seen in the fact that a solution has been chosen which is familiar to a PC user. The PC takes over the exchange of data and administration of the interfaces and at the same time the control station function. The MS-DOS operating system, however, is adapted as the version 3.XX can only manage two I/O ports (RS-232/Centronics).

The chosen solution can also be implemented without any problems from a hardware point of view because only two boards, each with two interfaces, have to be inserted into the slots and a connection established to the individual operating equipment. If one further assumes the normal training situation in which the individual operating equipment is connected to a PC or programming device, then by using an interface switcher it is possible to connect the learning station as well as the control station. This solution is also relatively simple as it only requires the interconnection of the various PC's.

5   Software Manager as the coordinator for FLS

Data communication based on the pedagogical concept of the FLS must ensure the interaction of the learning stations in a well-coordinated software environment by means of a Software Manager. This manager must guarantee the administration of the PC interfaces using a training-oriented (user-oriented) user interface. The S.A.A recommendations were chosen as the standard for the user interface of the coordination software. A hierarchical structuring of the various organisation levels was performed with pull-down menus for each of the subordinate functions. The levels can be looked upon as a reproduction of the process control chain which can be used for the layout of the heading in the individual screens within the Software Manager.

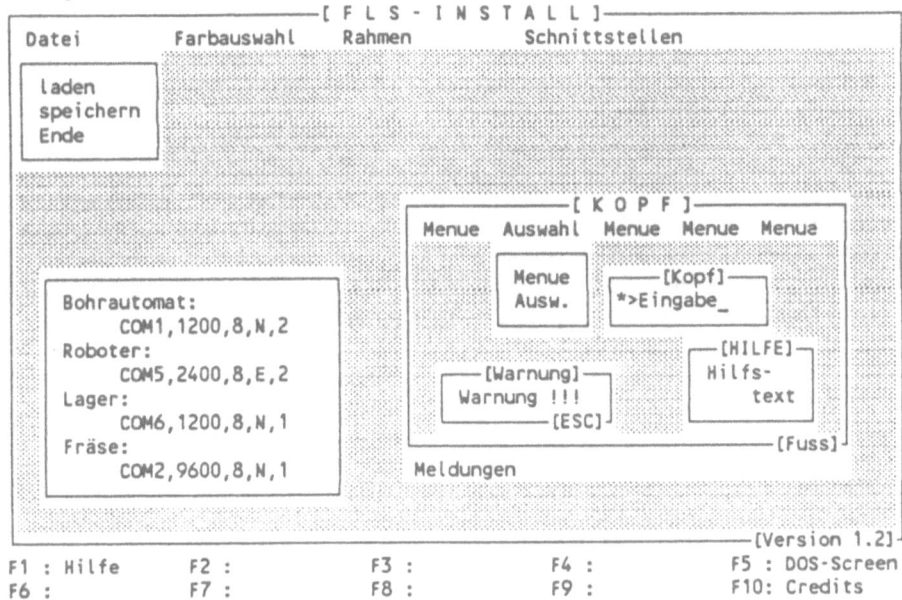

Fig. 2   Screen munue in FLS software

As a rule, programming software provided by the manufacturer is available for the various operating equipment (learning station). This software is also used in the FLS concept in order to program the individual operating equipment. It was inevitably necessary to have direct access to the software packages of the firms FESTO, AEG and IAET. Through the integration of the programming software, which can be called from the manager, these packages can be used, as up to now, to program the operating equipment.

This integration of software obtained from industry justifies the claim that continuity has been created in vocational training concerning both guidelines and company aspects. The student and instructor are confronted with a familiar software environment they are familiar with due to previous qualifications. This frees resources which can be utilised for further tasks in the integration of system components.

In its concept and structure the FLS Software Manager is oriented to commercial software. Using a menu-driven installation program (FLS-Install), it is possible for the user to set up all the parameters which are required to integrate the learning stations. These parameters are then stored in an editable CNF file. At the same time these settings allow integration of the software for learning stations/operating equipment, whilst being assigned to windows, as well as configuration of the interfaces to the operating equipment. When the transmission parameters have been set up, communication between the individual control units of the learning stations is defined. This enables the exchange of data between the control station and the process.

Coordination of several operations is performed using the FLS Software Manager's batch processor by parameterising the operating equipment's programs to be run via predefined program inputs and outputs. The associated control of the process makes it now possible to connect the operations in a serial or parallel manner so that it is possible to model different forms of work.

The operation-oriented control commands for the managing of a job via the equipment are written using a text editor and are defined in plain language. It is also possible, via the FLS Software Manager, to access the executable programs (EXE) and the data of the process so as to integrate visualising programs, which, on the one hand, display the course of events and, on the other hand, show the load on the individual operating equipment by logging the operational data.

The software linked to the Software Manager can be split into three categories/levels. The lowest of these, used mainly for laboratory operations, is oriented to high-level and/or machine languages. A further category is technology-oriented and therefore tailor-made for a single functional area. The supervisory control system software, as the third category, was designed to be function/operation-oriented and therefore intended for job planning functions. This distinction between software categories makes it clear that there is a large target group for whom FLS can be employed for training, further education and vocational training.

The various target groups require knowledge transfer which is tailored to their individual needs. In order to satisfy this demand, it is ·possible to link computer-aided audio-visual systems for both instructional and learning purposes. Using the Software Manager, it is possible to activate film sequences provided with a frame code from both a video recorder as well as from a video disk. The possibilities thereby created allow flexible learning compared with the traditional training concepts. At the same time the software allows the instructor to create his own CBT and incorporate it into the learning process.

6   Possible uses of FLS during training of skilled workers

In order to show how key qualifications can be attained during lessons at technical colleges, we can look at the proposed curriculum for the state of North Rhine-Westphalia. However, these remarks also apply to technical college lessons in other German federal states.

Fach	Ausbildungsjahr			
	1	2	3	4
Wirtschaftslehre	40	40	40	20
Fertigungs - und Prüftechnik	80	40	80	0
Werkstofftechnik	40	40	40	0
Maschinen - und Gerätetechnik	40	40	60	80
Steuerungs - und Regelungstechnik	40	60	40	20
Informationstechnik	40	40	0	0
Techn. Kommunikation	80	60	60	40
Summe der Unterrichtsstunden	360	320	320	160

Fig. 3   Timetable of the main subjects for Industrial Mechanic/Production Technology

The timetable for the reorganised industrial metalwork occupations shows the number of lessons for the main subjects during the individual years of the course (Fig. 3). The topic of machine and device technology plays a special role because of the number of lessons and therefore meets the job requirements for an Industrial Mechanic specialising in Production Technology. The aim of the lessons in this subject area is to enable the students to be able to analyse the functional design of technical systems and explain the interaction between the functional units of a system. Therefore, the focal point of the lessons should not consist of dealing with individual machines and devices - they should be dealt with regarding their systems engineering aspects./8/

The methodological question concerning the attainment of key qualifications in technical colleges, which include thinking and acting in systems, is answered by REETZ with reference to the technical college approach to action-oriented learning

- by the combination of work and learning in projects,
- with the help of learning offices, case studies and planning games,
- using instructional texts.

Despite the problems with which technical colleges are confronted concerning the subject principle, examination system and head-on teaching, they are given a good chance of being able to cope better than before with more complex interdisciplinary qualifications.

The reorganisation of the metalwork occupations has created the possibility of leaving the pure subject principle. The proposed curriculum stresses that "the teaching of acting competence places special demands on the way lessons are organised and carried out". /7/ Special significance is attached to the teaching and/or the usage of working techniques as well as the realisation of different social and action forms to further independence, interaction and the ability to cooperate. The carrying out of a technical experiment and the elaboration of a project are viewed as suitable examples of this.

Model experiments have been carried out at various firms in the Federal Republic of Germany in cooperation with the Bundesinstitut für Berufsbildung (BIBB) [Federal Institute for Vocational Training] in order to further vocational training taking into consideration structural changes in the organisation of work. These experiments have looked at the question of the promotion of mobility, the development of social competence and autodidactics. The investigation resulted in vocational training models which were different in each firm but their foundation consisted of accepted key qualifications common to all firms.

Concerning the design of the teachware for projects which are to be realised using FLS, we refer to the project and transfer-oriented training (PETRA) from the firm SIEMENS /9/ in which the objective of the vocational training was changed to methodological emphasis on the instructional text supported project method, superceding the Four-Stage-Method. The project method is regarded as the central factor in the teaching of key qualifications as well as in the acquisition of the knowledge and skills required by a specific profession, with focus on the taxonomy levels, reorganisation, transfer and problem-solving as recommended by the educational council for the differentiation of levels of learning in 1970.

The main areas of key qualifications, identified in the model mentioned, are oriented to future working requirements, which themselves correlate with training for a computer-controlled factory. As the model shows experience with a project method based on instructional texts, it appeared sensible to base the teachware methodologically upon it and thereby orient the design of the learning process to the organisational forms of the PETRA model. However, considering the degree of complexity which the FLS exhibits as a media configuration it was necessary that it not only encompass analysis but also synthesis of interconnected systems.

A possible project task could consist of designing, installing, commissioning and keeping error-free a flexible manufacturing plant for the machining of cuboid pencil holders with a maximum edge length of 120 mm using a CNC milling cutter. Small groups should put each of the plant's subsystems into operation and the team should prepare the initial operation of the entire system.

This project is designed for the vocational training of Industrial Mechanics specialising in Production Technology during the third/fourth year of the course. Its concept is inter-disciplinary with the main part of the learning content related to the machine and device technology.

Assuming the media configuration shown in fig. 1, we can split the above project example into the following component tasks:

- Design of a manufacturing plant for the machining of a cuboid pencil holder

- Assembly of the subsystems according to the installation plans on hand

- Deduction of the demands placed on the control system between the subsystems from the tasks of the entire system

- Writing the SPC programs to control the subsystems, input of the programs into the programming devices, syntax test, correction of programming errors

- Initial operation of the subsystems: loading the program into the SPC, subsystem trial run, diagnosis and correction of errors in each subsystem

- Initial operation of the entire system (by the whole group) : execution of the programs from the entire system's programming device, trial run for entire system, diagnosis and correction of errors in entire system, continuous operation of entire system.

It is a pedagogical/didactic principle that a student is led to something new from something already familiar or learned. The familiar objects are the learning stations at which problems were solved during the basic training. The integration task and/or the project is new. Thus, FLS consistently follows its previous intention, for the integrated training and the development of teachware, of falling back on existing training modules developed by diverse media manufacturers or educational institutions. For motivational reasons, documents created by students and instructors were also included.

From a design point of view, the FLS is regarded as an open system which allows structural modelling of manufacturing processes which can, however, be represented by various complex projects. In order to provide for differential structuring of the learning process (see also prototyping) in accordance with prospective vocational training, /10/ the institute is creating a teachware framework based on instructional texts. This teachware consists of specialist information and a manual for setting up the FLS.

7   Integration of FLS into a learning environment

The job of integrating FLS into a learning environment depends most of all on the conditions encountered locally; the basis for any further interconnection is represented by the basic configuration with network concentrator. The top-level processor in this concept serves both the control, processing and transport layers and is therefore open for connection to further networking. Analogous to the information structure in a computer-integrated factory, potential learning stations, such as design, product planning and supervision, can be included in addition to those connected with production.

The networking concept shown in Fig. 1 displays the interconnection of the leading processor/cell processor with the control layer. The learning stations (partly prepared for CBT) have both a CAD workplace as well as a product planning system installed. As the network can transmit both executable programs (EXE) as well as data (ASCII code) and the processor's hard disk is available for data storage, it is easily possible to realise the required data transfer and access functions in the computer-integrated factory - for example, CAD-CAM interconnection using the respective network software. This also means that it is possible to structurally implement the idea of a training plant (learning factory). /11/

8   Connecting FLS to simulations

The media configuration shown here can only partly clarify the real relation to a workplace in a computer-integrated factory. This is especially valid if, for a particular workplace, training is desired which is supposed to go beyond the teaching of automation technology, which is regarded as an inter-vocational field of learning./12/ Therefore, in connection with lessons in technical colleges and elsewhere, the question arises as to which media aids can be used to become familiar with the technological processes during automated manufacturing and depict the flow of information. Simulation models must be granted a higher status in this matter as it is not possible for technical colleges to perform experiments on real systems for cost reasons. /13/

Today the use of simulations in various learning processes is pedagogically undisputed, especially if the range of such programs is not overestimated. /13/ In CNC and SPC technology we find examples of high-quality simulations which have been extended to include robot technology and even entire plants such as a paint shop, and they have been enlisted, as it were, as media for training. In order to be able to depict the manufacturing processes which refer to workplaces on computers in the learning station environment described here, one must use planning systems for factory layouts which run under MS-DOS and by means of simulation permit planning games which refer to concrete manufacturing situations. /re;sp/ By adapting the display interface to the onsite operating equipment, it is possible to convey planning action strategies in a creative and organised manner.

9   Summary

According to Taylor's dogma, industrial work implies little identification for the employee because of its low job content. This is intensified by the fact that, due to increasing automation, this content loses out to the simultaneously increasing tendency towards rationalisation. If work should also form a purpose in life for employees' personal concepts, then the idea of moving away from performance production towards an object-oriented arrangement is, on the one hand, pioneering, but it requires different concepts during vocational training. At the same time, it also appears to be necessary to regard the starting point, namely automation, as being not only technocentric but also shapeable in the anthropocentric meaning. The FLS has attempted to take up this idea by fulfilling the prerequisites for designing flexible manufacturing systems based on existing media in a product-oriented manner. Since, in the project task described here, the FLS is to be generated under planning aspects and with an integral understanding, the apprentice experiences the working world as changeable. This also leads to the elimination of any fears of the use of flexible automated manufacturing systems.

Since the subject content is learned according to set rules using the media to be integrated, it is possible that this content, as an experience potential for the group, contributes to the success during the design of the structures in the computer-integrated factory if the phases of the product manufacture are considered as a whole. The significance of utilising experience and designing process interconnections represent a special status for learning in networked systems. Organisation of the media configuration and decision-making in a project which, because of its complexity, can only be performed by a group and possibly only by the instructor, can only be completed successfully if all those involved behave in a positive social manner. This is also true if a predefined media configuration is chosen as shown in the sample configuration since setting up, operation and maintenance in a project task based on equipment can only be carried out by a group.

If, during pedagogical/didactic discussion, an orientation towards key qualifications is carried out de facto despite academic doubts /17/, then the development of the transfer capability is also affected. Transfer means that the systemic acting and thinking is taught by means of the concept used here, which facilitates the initial training and acting, such as in a processing cell. At this point, special attention must be paid to the results of the teaching of knowledge concerning CNC technology using both model and original machinery. It was possible to use the former to attain 75% of the learning content, /14/ which reinforces the significance of the equipment for CIM learning processes. Moreover, if one considers the cost situation, then one must assume approx. DM 2000 for the Software Manager. The cost of the interconnection of the learning stations shown in Fig. 1 would be in the area of DM 300,000. The training path shown here is also interesting if one considers the training costs themselves.

Attempts at the implementation of similar concepts can be found in both relevant literature and at several companies. During the next year, the institute will carry out an empirical investigation of this concept, in order to present, in detail, results concerning its pedagogical efficiency. In the future, one can assume problem-free implementation due to the standardisation of interfaces (protocols), on the one hand, and software for programming automation equipment during the next few years, on the other. Along with the FLS this will create the hardware and software prerequisites for the interconnection of learning stations.

The inclusion of new contents and methodological concepts into the guidelines for vocational training inevitably results in the necessity to develop new media and to employ different teaching methods in technical colleges. The concept of a Flexible Learning Laboratory System (FLS), as presented here, attempts to make a contribution to this goal.

10   Literature

/1/   Bullinger,H-J.; Kern,P.: Arbeitsgestaltung in hoch automatisierten Produktionssystemen. In: Elektrotechnische Zeitschrift-ETZ 112 Jg H13/14 91 S. 676-680

/2/   Meyer-Dohm, P.: Bildungsarbeit im lernenden Unternehmen. In: Meyer-Dohm, P.; Tuchtfeldt, E.; Wesner, E.: Der Mensch im Unternehmen. Bern, Stuttgart: Paul Haupt-Verlag 1988, S. 249-271

/3/   Verordnung über die Berufsausbildung in den industriellen Metallberufen. 1987, S. 24

/4/   Sonntag, Kh.; Benedix, J.; Heun, D.: Kognitive Anforderungen bei Anlagenführer- und Innstandhaltungstätigkeiten. Zeitschrift für Arbeitswissenschaft. 1991 43. Jg. H1 S. 26-33

/5/   Ulrich, H.; Probst, G.: Anleitung zum ganzheitlichen Denken und Handeln. Bern, Stuttgart: Paul Haupt-Verlag 1988

/6/   Reetz, L.: Zum Konzept der Schlüsselqualifikationen in der Berufsbildung. In: Berufsbildung in Wissenschaft und Praxis, 1989, H5 und H6.

/7/   Bracht,F.: Produktionsnahe Ausbidung an automatisierten Systemen. In:lernen und lehren. Bremen 1991 6.Jg H22 S. 42-50

/8/   Der Kultusminister des Landes Nordrhein-Westfalen: Neuordnung der industriellen Metallberufe. Vorläufiger Lehrplan für die Berufsschule (Lehrplanentwurf Mai 1987). Düsseldorf 1987

/9/   Boretty, R.; Fink, R. u.a.: PETRA - Projekt- und Transferorientierte Ausbildung. Berlin, München 1988

/10/   Rauner, F.: Anmerkungen zu einer prosektiven Berufsbildung für die "neue Fabrik". In: Laur-Ernst, U. (Hrsg.): Neue Fabrikstrukturen - veränderte Qualifikationen. Berlin: BIBB 1990, S. 51 ff.

/11/   Behrens, A.: Lernfabriken sollen CIM-Idee verwirklichen. Technische Innovation und berufliche Bildung. 1987, H. 4, S. 22 ff.

/12/   Klinger, H.:Konturen eines Lernfeldes. In: lernen und lehren. Bremen 1991 6. Jg H22 S. 11-22

/13/   Kiel, K.H.; Komoll, J.; Langmann, H.-W.: Medien für eine qualifizierte Arbeit in Fertigungszellen und Fertigungssystemen. In: Hoppe, M.; Erbe, H.H.: Rechnergestützte Facharbeit. Wetzlar 1986

/14/   Laur-Ernst, U.: Mit CNC-Simulatoren lernen - Möglichkeiten und Grenzen. In: Hoppe, M.; Erbe, H.H.: Rechnergestützte Facharbeit. Wetzlar 1986, S. 106 f.

/15/   Reinhard, A.:Integrierte Methoden als Basis integrierter Fabrikplanung und -steuerung. Zeitschrift für Logistik. 8 (1987) Nr. 10, S. 44 ff.

/16/   Springer, G. (1991). Qualifizierte Mitarbeiter - ein Schlüssel für die integrierte Auftragsabwicklung. Vortrag auf der VDI-Fachtagung "Integrierte Auftragsabwicklung" am 20./21.06.91 in Nürnberg

/17/   Lennartz, D.: Thesen zu Schlüsselqualifikationen und Qualifizierungskonzepte. In: Laur-Ernst, U.: Neue Fabrikstrukturen - veränderte Qualifikationen. Berlin: BIBB 1990, S. 103

CIM Training for open and migrationable technology

Prof. Dr. Wolf Martin
University Hamburg
Hamburg Germany

The basic idea behind designing an open and migrationable learning centre for CIM training is that today CIM is a development strategy for future factories and not a complete technical concept. It cannot be expected that the numerous current development trends will lead to the generally accepted uniform concept of the CIM factory. It is much more likely that in the future a variety of factory and production-specific manufacturing methods will continue to determine the development of factories in the years to come.

Uniform organisation and data structures in the whole factory are necessary for a computer-integrated production which should cover all working-areas from receiving orders, design and production up to delivery. Therefore, the path leading to CIM in the factory primarily consists of a development of organisational and working structures and only secondarily involves the installation of new technical resources.
There is no sense in holding on to current factory structures - even if they may be efficient for the time being - without questioning their meaning for future development. If these structures are implemented in the technical system, the basic aim of the CIM strategy might be blocked for years if not for decades; increasing the flexibility of production would only be possible at lower levels of production and would take place at the expense of flexibility for the whole factory.

Therefore, CIM does not represent a certain technical solution for flexible automated fabrication, rather it describes the idea of adaptable factories, which consistently make use of modern information and communication technology to meet changing market demands. Consequently CIM as a development strategy is dependent on open and migrationable organisation models, open and migrationable technical components, and on highly qualified flexible employees.

For this reason CIM training requires an open and migrationable vocational training system. In the current situation it would be a mistake to develop vocational training systems irrespective of the practice in factories or to generally implement and establish specific CIM solutions. Because, on the one hand, a final result of CIM development cannot be anticipated and, on the other hand, prospective training is becoming more and more important, only open and migrationable equipment concepts that consider the current patterns of development for CIM as a whole can be regarded as forward-looking.
This openness of the vocational training system implies additional effort during its

installation but only in this way is it possible to guarantee the necessary flexibility for years to come. In view of the outlined dynamic development in factory practice, vocational training must be arranged and organised in a matter that recognises the necessity of appropriate experimentation and development surroundings for CIM training; here inflexible complete solutions would be as out of place as in factory practice.

Nevertheless, training in the described surroundings requires that preliminary decisions covering curriculum, educational subjects, and technical equipment are made today. The relatively open future of CIM development calls for open instruction concepts and open modular vocational training systems. These vocational training systems should allow gradual expansion of computer-aided technologies, i.e. realising single CIM learning islands and integrating existing training concepts and systems.

The primary aim of imparting prospective vocational competence is to make the trainee capable of taking part in the organisation of CIM development in factory practice and to impart specialised knowledge about the respective current stage of development. The development of organisational structures and work arrangements is an important component of CIM development and must therefore be integrated as an essential element of training.

In the current situation it is necessary to clarify subject areas and the corresponding technical equipment needed for a start towards CIM training which enables trainees to actively and competently take part in future CIM development.

With regard to CIM as an organisational innovation strategy, the following areas of interest are significant for vocational training and thus for the development of an open and migrationable learning centre:

- development of organisational and working structures,
  management processes and planning instruments
  forms of work organisation and workplace design,
  flow of information (horizontally and vertically) in the factory
  data structure and stock of the factory
  personnel, factory, and machine data capture

- development of technical production systems
  CIM concepts and systems
  CIM components (CAM, CAD, CAP, CAQ)
  integration of CIM components (computer networks, data flow)
  computer-controlled units and machines
  interconnection of units and machines (interfaces, bus-systems)
  configuration of computers and networks
  user interfaces and programming systems.

In the course of vocational training the interdependencies between these areas must be elaborated; on the one hand, certain organisational and working structures require a certain form of technology and, on the other hand, a certain technology (e.g. network configuration) sets limits to the free play in development of organisational and working structures.

In each of the areas of interest knowledge about the historical genesis of the current form is important to point out alternatives and future developments.

The following examples of realisation of CIM vocational training centres (SD 2002 CIM Training System, CIM factory Hannover, FESTO/IBM learning system) will show that the training systems focus on technical production systems.

The SD 2002 system claims to make training possible at all factory levels, from the process level to the management level. A modular system is implemented at the first level. A flexible fabrication cell can be configured consisting of real machines and units, such as CNC machines, robots, transportation systems and measurement stations. Through the cell network all system data is available at the control terminal and can be used for production control. Even though the system is basically open (on the basis of industrial standards), as yet additional interfaces are not available at the machine and unit level; the integration of extrinsic elements is only conditionally possible. Basing the fabrication cell on a factory network, for example, on a CAD or programming area, seems uncomplicated (Novell network). Simulation models are supposed to expand the CIM island and provide for additional experimentation on the implemented CIM island.
The described system primarily refers to the CIM components of the technical production system. Through the use of machines, units, and networks in factory practice a basic openness is provided at the CAD/CAM level, which would still have to be specified and developed further.

The CIM factory in Hannover serves research and development of technical components as well as CIM training at universities and vocational training. The factory is made up of three model companies, two of which have real production resources while the third exists as a simulation model. The three companies are networked so that demonstration of the material and information flow between them during completion of orders is possible. Training is focused on the technical CIM component, i.e. the computer-aided technologies; vocational training only plays a small part. The factory is equipped with industrial components from different manufacturers so that the technical structure is basically open. The majority of interface problems at the level of machines and units seem to be solved. As far as research is concerned, the CIM factory is working on the technical realisation of organisational structures, not in the sense of developing organisational structures within in the scope of CIM, but of integrating CIM components in existing structures with the help of integrated company database systems. Here workplace structures and work arrangements make no difference. The research area concerning migration support of various computer-aided technologies offers interesting aspects, which would still have to be realised in vocational training.

In contrast to the first two examples of CIM training centres based on industrial machines and units that - conditionally - make real production possible, the FESTO/IBM system is equipped with teaching models. Established on the basis of the FESTO's "modular production system" with the four stations of stockkeeping/handling, stockkeeping/assembly, stockkeeping/assembly/welding, and quality control, a fabrication line for toothed gears has been realised as a model. The communication and network control is implemented on the basis of IBM industrial hardware and software. The teaching system focuses on the production technology (CAP) but thanks to the IBM software, it is also open for further CIM components (CAD, CAQ, PPC).
Integrating additional industrial machines and units into the "modular production

system" is hardly possible because of the special FESTO controlling system. The extent to which the teaching system can serve as one component in a superior equipment concept must be researched more extensively. A high degree of modularity and flexibility exists at the level of mechanical unit and machine models.

All of the presented examples represent more or less technically open teaching systems for computer-integrated production, i.e. for CAP and CAM. Regarding the stage of CIM development, they are equipped with interfaces to further CIM components. At the presented machine and unit level they are implemented in a modular and flexible way, and represent development and experimentation surroundings for this level. Integrating extrinsic machines and units is basically possible; however, the construction of interfaces in actual cases might be complicated and call for considerate effort. In these cases systems equipped with industrial machines and units are probably more suitable.

The learning systems that are essentially made up of a flexible fabrication cell or line do not yet offer ways of pointing out subjects of interest superior to technical production, such as organisational structures and development of working structures. With respect to the equipment and training concepts, they only represent a part of CIM development.
These subjects of interest are the ones most complicated to be transferred to the learning centre. One way to remedy this problem is simulation at a level beyond the CAD/CAM field: the factory level. Despite all restrictions associated with simulation models, combining real CIM components with simulated ones makes it possible to demonstrate CIM development in a factory as a whole, including aspects of management and production in the learning centre.

The examples illustrate remarkable partial solutions on the way to a open and migrationable learning centre, even if implementing them may require considerable effort. Especially the development and use of training concepts in vocational training practice must not fall behind technological development in the learning centre.

Training facilities for CIM

Flavio Canetti, Reinhold Dreßler
PERFO/Siemens
St. Croix Switzerland/Nürnberg Germany

1 Introduction

Being competitive and continuing to compete effectively represent a major challenge for today's manufacturing enterprise. In order to address this challenge, the enterprise must find new ways to lower product costs, improve product quality, reduce inventory, shorten lead times and respond more quickly to customer and market demands on a global basis.

The increasingly competitive manufacturing environment is forcing enterprises to re-evaluate day-to-day operations and in doing so, they are discovering the need for a comprehensive, integrated, enterprise-wide solution for now and for the future. They need systems that enable different departments to work more closely together that allow engineers to take automated shop floor equipment into account when designing their products and that extend communication lines to suppliers and customers as well as to other internal departments.

Such systems exist today, although they are perhaps only partially integrated and are not implemented at all levels. They are, however, all based on a new strategy or philosophy

C I M

or Computer Integrated Manufacturing, which harnesses information system technology to integrate all manufacturing, commercial and administration objectives. When implemented correctly, it

- increases productivity
- increases quality
- increases cost efficiency

and encourages responsiveness throughout the enterprise.

2 The definition of CIM systems

CIM systems are referred to as any computer-oriented equipment or system which aids in or effects automation of a manufacturing enterprise and which is planned to increase, if not eventually complete, integration of the enterprise.

CIM typically involves

- product design
- process design
- process planning
- process control
- production support
- decision support

3   The real aim of CIM

As a consequence of ever more sophisticated customer demands for

- greater variety of products along with
- quality
- reliability
- Innovation

strategic business goals have changed through the years.

CIM is the answer to four key trends in industrial society:

- global markets
- demassification
- shorter product life cycles
- complex products

The real aim of CIM is

FLEXIBILITY

By means of CIM, the cost aspect of product diversity can be reduced. At an internal company level, the traditional conflict between marketing (which wants to offer customers more models) and the factory (which has wanted to limit product line variety for the sake of product efficiency) is diminishing with the implementation of CIM.

The global objective of "more flexibility" broken down to the operational level means: "reduce order lead time" and "reduce work in progress". In fact, these are the main goals of shop managers today. Decrease throughput time instead of cost.

4   Obstacles to the implementation of CIM

One factor delaying the implementation of CIM is a great gap in economic thinking. Sensible evaluation of advanced technologies must take into account quality, flexibility, competitiveness and productivity. Today, competitive ability is of far more importance than short-term costs.

CIM technology investment effects are long-term, and therefore difficult to assess. At this level, there is still a need for efficient cost evaluation methods.

However, by far the most important overriding obstacle to the implementation of CIM is

THE LACK OF QUALIFIED STAFF

at all levels, from operator to engineer to project manager to top-level management.

Studies have shown that out of 100 % obstacles

-	lack of training accounts for	25 %
-	financial strength	20 %
-	profitability	20 %
-	technological shortcomings	15 %
-	entrepreneurial shortcomings	15 %
-	technology hostility	5 %

5   CIM = a must and a chance

For large as well as for small and medium-size enterprises, the proliferation and support of CIM as the logical extension of the traditionally well recognized manufacturing equipment business of Europe is a major task for now and for the future, for managers, politicians and institutions to avoid further erosion of competence in this vital area of the world.

6   Training

To gain more acceptance, however, the prejudice against CIM has to be overcome. And how can this better be achieved than through Training, right from the start of a career followed up by Further Training and Re-Training.

However, Traditional Educational Methods are not Applicable in a Volatile Technology Environment i.e. training on the job is obviously not acceptable, as this would interrupt the production, with consequences to costs and quality.
The task therefore has to be transferred to a Training centre.

Prior to the planning of such a training centre the objectives and criteria have to be established:

- Clear Objectives which are in Line with Long-Term Policy
- Objectives which are Based on Existing Systems (no Revolution!)
- Objectives which Meet Existing Requirements
- Objectives which are Progressively Implemented

On the basis of these requirements, the criteria for the equipment, software and teachware to be used for training are:

- Equipment which Effectively Reproduces Real Industrial Conditions (No Toys!)
- Equipment which is adaptable and Allows for Evaluation
- Equipment which is Safe and gives Students Confidence
- Equipment which Provides a Degree of Economical Profitability (used for Production)

With training programmes using this philosophy, the student will encounter similar conditions and systems in the factory after completing his training course and will therefore adapt to his new task more easily.
The continuous development of systems in industry can be included in the training centre without interface problems by using industrial type equipment and software. Therefore, training must be industry-oriented.

7 Training Systems

Let us view first the hierarchical structure of a company, starting with the process and finishing at corporate management, all the steps have to be incorporated in a successful training programme.

To be efficient, it must be possible to train each section individually, starting from the foundation and continuing right through to the roof.

To achieve this goal a concept has been created to start from training islands such as PLC, CNC, robots, etc. through bus systems to flexible manufacturing cells, ultimately leading to the complete CIM Laboratory.

The basic concept of the training islands right through to the CIM laboratory is to use an industrial product, modify it with didactic aspects for hardware, software and teachware to create an efficient training setup.

To summarise, the development of the training programme is followed by

- Creating autonomous automation unit workplaces (modular system)
- Preparing models (simulation) for machine tools, transfer lines, etc.)
- Defining and preparing training-oriented user software, applying the existing basic software of the various components
- Combining several workplaces to form a practical production unit, e.g. flexible manufacturing cell and CIM Laboratory
- Generating data pools for simulating different operating conditions

The islands are interconnected to form a laboratory for industrial automation and eventually to reach the ultimate goal of a complete CIM Laboratory, simulating a real computer-operated company.

8   Education and training in the CIM context

Although often ignored or neglected, education and training are important parts of the CIM strategy. A CIM concept introduces major changes in job content and in organisational structure throughout the enterprise. Without an adequately trained workforce, without its understanding and acceptance of what is involved, a CIM concept can never get off the ground. This calls for a collaborative effort between the agents concerned, i.e. industry and education/training and research/development institutions.
The latter have to understand that industry needs a better-motivated, more flexible, multiskilled and self-disciplined workforce that needs to perform as an individual and as a team member, as a vital part of a larger system. There is, therefore, a need for innovative solutions to training problems. There is a need for a more systemic and system-oriented approach.

To meet these new.requirements, training objectives need to be practice-oriented so as not to widen the existing "knowledge and competence" gap between industry and training institutions. By the same token, training systems and equipment should be able to reproduce real production conditions as closely as possible.

9   The case for realtime simulation

In education, simulation is used to duplicate activities which are usually too demanding of resources to allow their actual implementation, whereas in a professional environment, simulation is a design or planning activity in preparation for an actual physical process.

However, simulation contributes more than mere financial savings to the education process. Many real world situations are too complex or too dangerous to allow student experimentation. Moreover, simulation offers a work environment free of stress, in which the student may makes mistakes and to observe their results.

There are levels of technology where the teaching process requires an exact replica of the real system in order to appreciate the physical and conceptual factors involved. A typical example is a control panel for industrial equipment.

Experience with simulation and realtime training equipment has shown that there is a need to strike a balance between abstraction and real design.
Specifically for the learning process in the engineering field, there is a need for students to actually "cut metal" before entering a workplace situation.
Simulation activities provide the basis upon which students are able to cope with real problems, but without an appreciation of what the real problems are, the simulation process lacks focus and direction.

In short, training systems should above all present the following advantages:

-        students should have contact with the real situation students should not "play" and substitute random trial and error for scientific method and

understanding of the principles involved
- training systems should incorporate multidisciplinary knowledge, even if the total of that knowledge is beyond the capability of a single individual
- students should be able to explore a variety of solutions within a reasonable time scale, and at a speed which discourages reliance on manual operations to arrive at an acceptable solution
- training equipment should be technologically evolutive, in order to meet requirements at all levels and to optimise investment
- in line with evolution in a wide range of current systems, training systems should move towards object-orientation
- training systems should also give students an understanding of the questions to which the technologies demonstrated are the answers
- particularly in the field of CIM, training should provide an understanding of manufacturing as being an organised system of basic (invariant) functions
- CIM should be seen as a strategy from which the basic manufacturing functions may be organised and supported and from which technological solutions may be derived.

10   The SD2002 CIM training system

Designed to meet training requirements within the framework of a classical CIM architecture (see below), the SD2002 system is integrated, modular, comprehensive and technologically evolutive. It covers the majority of training needs from basic vocational training to higher technical education.
These range from CNC programming and simulation, machining, robotics, etc., to more complex techniques such as cell control, handling, and integration.

At a physical level, the system consists of a flexible production cell incorporating CNC machines (milling and turning), robots, PLCs, etc., with an automatic guided vehicle to transfer workpieces and tools within the cell.
The network link and cell control software provide access to all system data for realtime production control, tool managements, material flow control, cell simulation, statistics, optimisation of cell work load, interactive maintenance planning, etc.

The specific system design provides ideal hardware and software support for training at different hierarchical levels, for example:

- management
- CAD system operators
- production planning personnel
- shop floor personnel
- maintenance staff

where objectives range from strategy to problems such as the influence of flexible production on design, the introduction of new flexible production parameters on the shop floor, etc. As the system design is open, it lends itself to further development, which is particularly interesting for research and development at the university level.

The SD2002 is presented in detail below (Fig. 1).

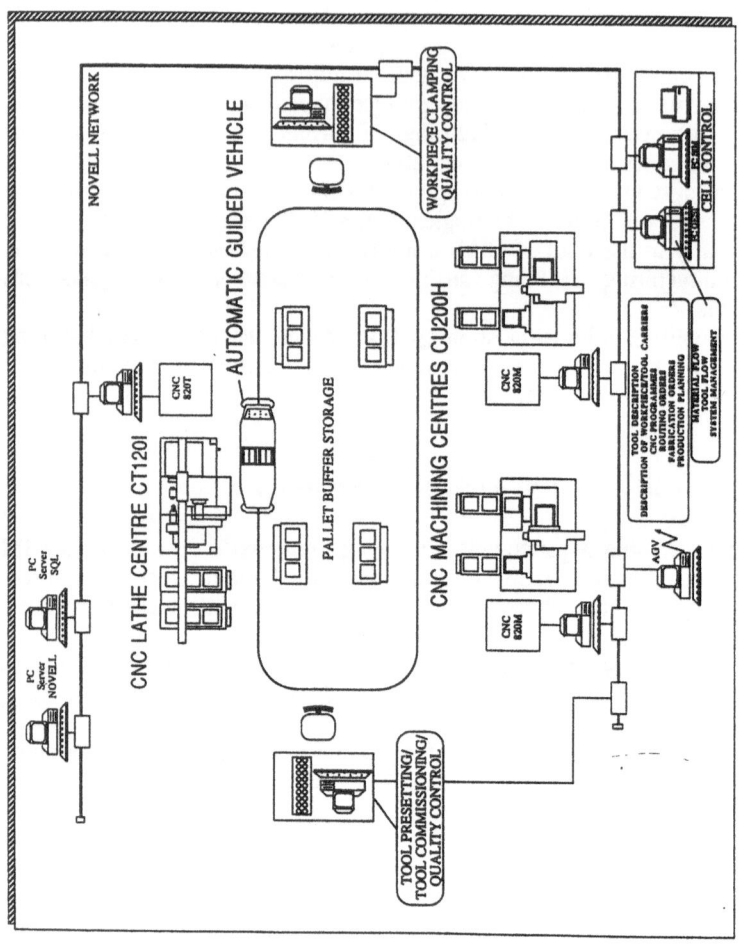

Fig. 1   Configuration of the SD2002 CIM Training System

CIM Training through CIM Practice

Frank Hanewinkel
CIM-Fabrik Hannover, University Hannover
Hannover Germany

1 Presentation of the CIM-Fabrik Hannover GmbH (CFH)

The CIM-Fabrik (CIM Factory) Hannover GmbH has been in existence since early 1988. It was founded by three professors: Tönshoff (Institute for Production Engineering and Machine Tools, IFW), Doege (Institute for Forming Machines, IFUM) and Wiendahl (Institute for Production Plant Engineering, IFA) of the University of Hannover. The foundation is meant as an institution for technology transfer (Figure 1).

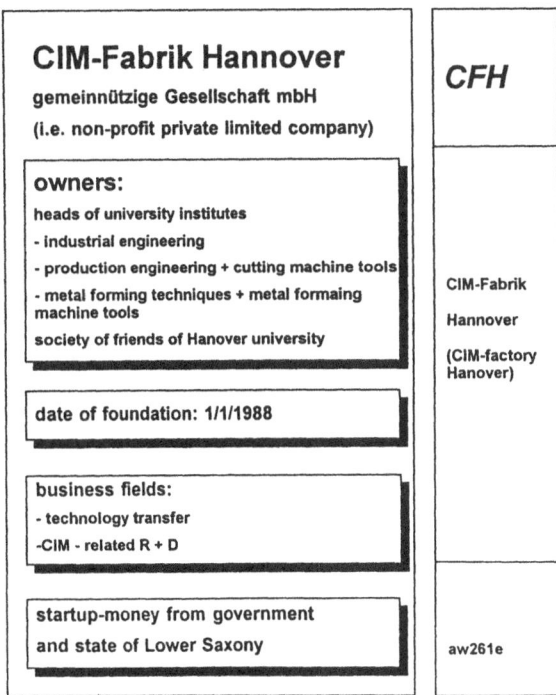

Fig. 1   CFH History and Purpose

Ever since the CIM-Fabrik has become a partner for small and medium-size enterprises (SME). For better performance of its assignments the construction of a new building was started in July 1990. In September 1991, after 15 months of construction the company moved into the new building, which officially opened on 11.11.1991 with a symposium [NN88].

In addition to participation in computer-aided production and industry research projects on a national and international scale, the CFH also has to act as a CIM technology transfer centre. In this concept education plays an important role (Figure 2).

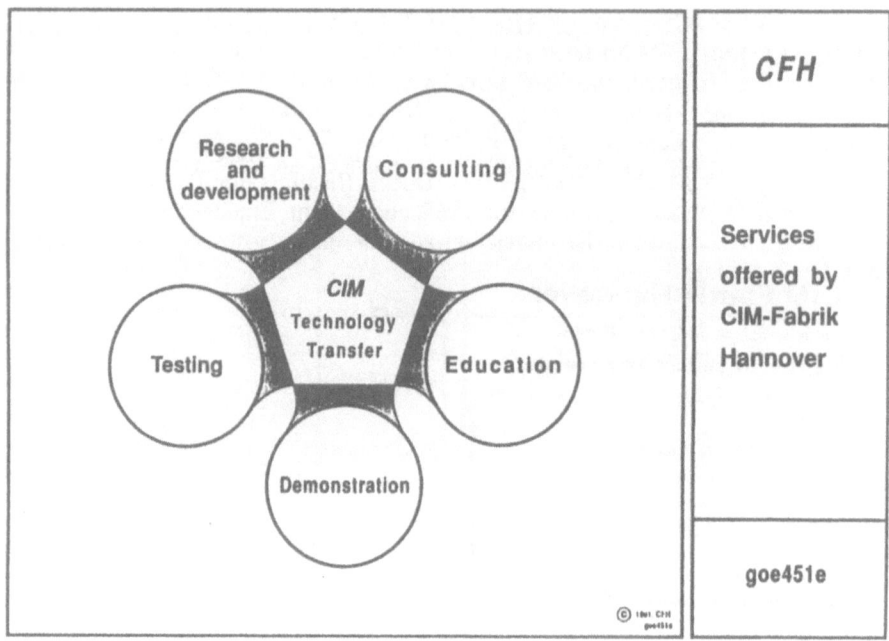

Fig. 2   CIM technology transfer

2   CIM education at the CFH

Education at the CFH is based on two pillars: university doctrine and company internal education with seminars, workshops and courses (Figure 3).

2.1   Students education at the CFH

An agreement of cooperation links the CFH to the University of Hannover [NN89]. As part of this cooperation the CFH carries out student education for the subjects

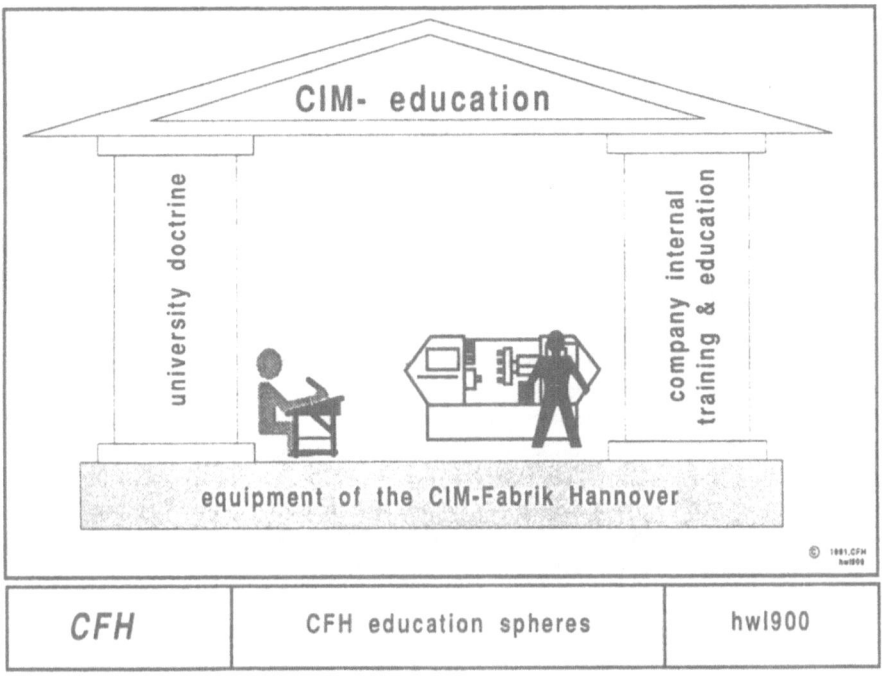

Fig. 3   CFH education spheres

CAD/CAPP and CAM/CAQC. Subject matters are:

CAD/CAPP

wireframe, surface and volume model,
CAD hardware,
modelling methods,
programming of CAD (API),
fundamentals of manufacturing planning,
computer-aided methods for operations scheduling,

CAM/CAQ

basics of organisation,
basics of data processing,
NC programming / WOP,
machinery programming with APT 1/2
control station technique
production data input (PDR);DNC;QDE
CAQC

The topic CAQC accounts for roughly 30% of the entire time of the course.

For both subjects exercises (CAD, NC programming, robot programming, manufacturing planning, quality control, etc.) are offered which are mandatory to achieve the correspondent training. Through these exercises the students come in contact with specific systems and actual problems the same as they occur in industry.

In addition to the subjects mentioned, the CFH also offers exercises for other lessons held at the University of Hannover. The subjects include production plant operation and PPC.

The assignments of the CFH within the University of Hannover were initiated by a cooperative agreement between the CFH, the university and IBM-Germany GmbH. One of the objectives of this cooperation is to promote CIM technology within university teaching. The students are offered the opportunity to achieve a certificate demonstrating their additional CIM training. Five out of twenty possible subjects are needed to achieve the certificate. Within the range of topics a special focus is on five fundamental subjects. These are CAD/CAPP, CAM/CAQC, CIM technology and economics, PPC and a programming language.

2.2   Company internal training and extended education

The second pillar of the CIM training at the CFH gives the industry a possibility of employee training and education by the CFH. For this target group seminars, workshops and courses are offered (Figure 4). Seminars are periodically offered and cover a different subject each time. Depending on the topic, the target groups vary from employees in leading positions (manager, department manager) to the operative level (master, foreman). The seminars treat specific problems connected with CIM as well as general aspects and questions about CIM planning and installation. A seminar is divided into a theoretical part, realised through lectures, and a demonstration specific to the subject. The demonstration is carried out by the use of the manifold CFH options.

During so called in-house seminars the offered subject may be taught and demonstrated directly at the company's site. In this case the participants are mostly employees belonging to the same control level who have to be comprehensively informed about a specific topic.

In-house seminars offer the possibility to adapt best the problems and subjects according to the company's needs. Thus companies may train their employees very precisely for a specific problem and possibilities to solve it.

The CFH education offered also includes workshops. Like seminars they have a theoretical element but practice with CIM components like PPC, control station or CAD is much more emphasised than it is in seminars, where theory plays the most important role.

Furthermore, the CFH offers courses treating CAD or quality assurance which are not listed in a general index. As they depend on an individual agreement with the

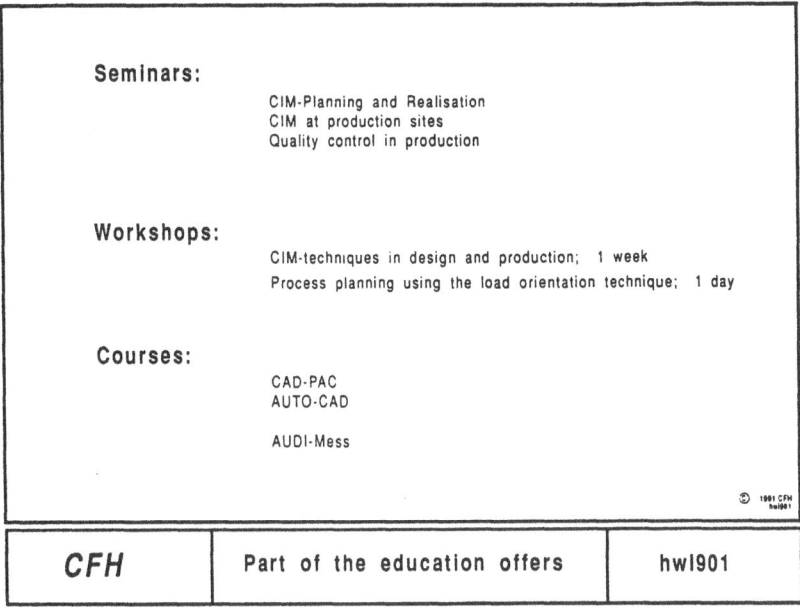

Fig. 4   Company training at the CFH

client they are more like in-house seminars, only a group of employees belonging to one company attend them. The courses deal with education and introduction for a specific software system. During these courses, which take between one or two weeks depending on the subject, the participants get to know the necessary fundamental software specific knowledge and how to operate the system while solving concrete problems.

3   The CFH demonstration area

Training and extended education at the CFH is integrated into a general concept that can demonstrate realisation of CIM in a company as well as cooperation of different companies through information technology [Tön91]. This demonstration concept is based on three model firms (Figure 5).

The model 'propeller shaft' company initiates an order concerning its whole system. Complying with its clients instructions it develops a new or modified propeller shaft. Some parts of the new shaft are wrought-iron work which is ordered at the 'forging shop'. The geometry of these parts is transmitted to the forging shop together with the order using remote data transfer. As the forging shop cannot manufacture the needed dies itself, it places an order at the third model firm, the tool manufacturer.

136

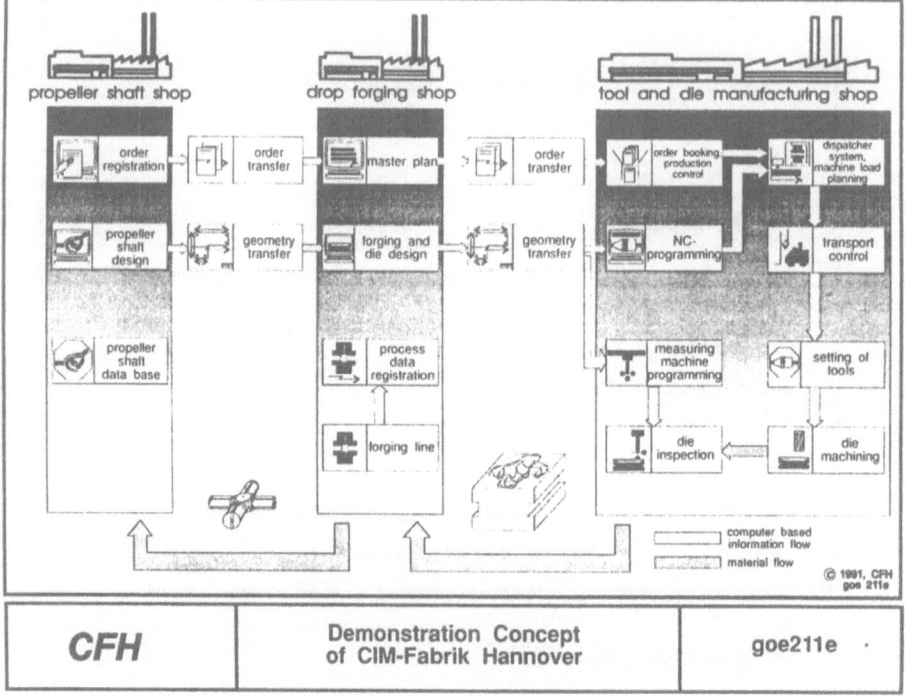

| **CFH** | **Demonstration Concept of CIM-Fabrik Hannover** | **goe211e** |

Fig. 5   CFH model company concept

For this order geometric data is transmitted, allowing the tool manufacturer to develop a production-compatible design of the dies and corresponding NC programming. Dies produced with the developed programs are delivered to the forge which uses them to manufacture the ordered parts.

In addition to this transmission of orders and geometric data, more and more emphasis is made on quality assurance.

The model 'forging shop' and 'tool production' firms within the CFH are not only represented on a computer but also possess real production facilities for manufacturing the dies and wrought-iron parts. This makes it possible to perceive problems of real production and to discuss and demonstrate them during an education.

The forge consist of a modern forge line (Figure 6). The beginning of the materials flow is marked by a billet shears (HKS 300, EUMUCO), cutting down rods of 6 m length to smaller raw material pieces. A spiral conveyor and vibrating conveyor transport the rod pieces to an inductive block heater (Type EBS, ABB) to reach forging temperatures. After passing the temperature controlled good/bad part switch, a robot (IR381/8.0, Kuka) picks up the raw material pieces and places them at the first step of the forming tool. The robot controls the forging press (screw press PSH 4.26f, Weingarten). It carries out the transportation between different forming steps and supplies the forged part to a fin cutting press. Finally the workpiece is placed on a cooling line. Here the workpiece is specifically cooled down.

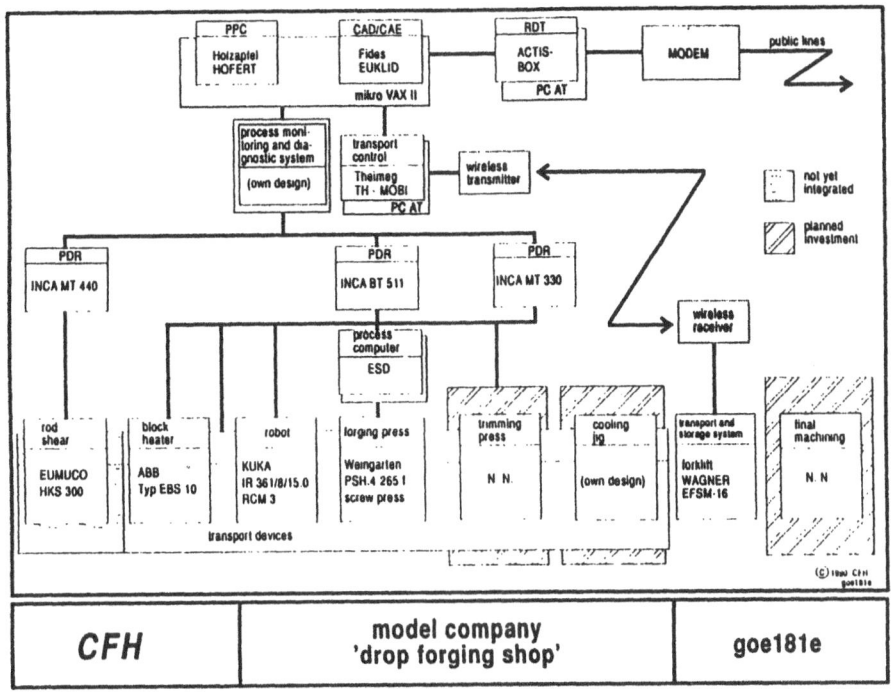

Fig. 6 Configuration of the model "forging shop"

Transport and stocking of rod material and finished products is done by a forklift truck that receives transport orders through radio transmission of data from the transport control station.

Shop floor data and process data of the forging line are collected with PDR devices (INCA) and are reported to the PPC system (HOFERT, Holzapfel). The design of the forging part to manufacture as well as the die surface is done by CAD (EUKLID, Fides). The geometry data of the final propeller shaft can be applied using a standard data interchange format (VDAFS).

The tool manufacture (Figure 7) possesses a machining centre for milling operations (DC 30, Controller Bosch CC300M, Deckel). A cavity-sinking EDM machine (DE 10C, Deckel) is installed for fine machining. The machinery tools for all milling machines are preset using CNC presetting equipment (H4123, Zoller). The measuring of all kinds of parts is possible by means of a 3D-coordinate measuring machine. This machine is used especially for the measuring of manufactured or worn forming tools. The programming is done with the programming system Quindos (Leitz). Further on a 5-axis milling machine (FP 2H, Deckel) is installed. All machines are connected by a shop floor LAN (DNC 6000, rwt) permitting the transmission of NC programs. The purchase and installation of a lathe centre and a transport and stock system are planned.

For the scheduling of shop floor orders a control station (AHP) is used. The correspondent production orders are generated by the PPC system (MAPICS-DB,

138

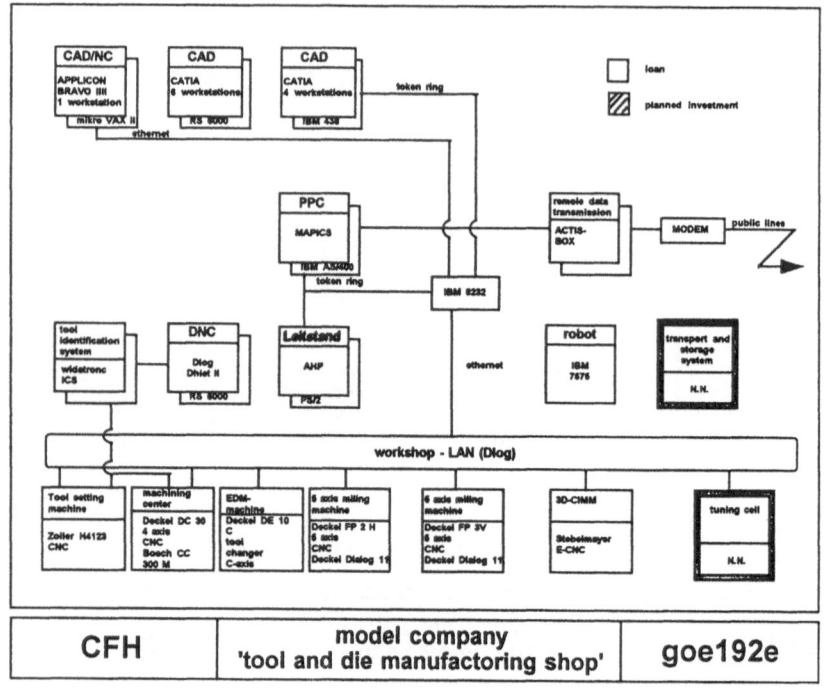

Fig. 7 Configuration of the model "tool manufacture" firm

IBM). Design and NC programming may be carried out using two different systems (BRAVO III, Applicon; CATIA, Dassault).

As a result a strongly heterogeneous computer and software environment is installed at the CFH. Mostly IBM and DEC computers are used, a constellation which can be found in various small and medium-size businesses. With good reason tool and appliance production were given special emphasis when the model firms were chosen. Typical firms of this industrial sector are small and medium-size businesses playing a crucial role as a supplier for the whole panoply of machinery and car production. The cooperation of those companies with their suppliers is of great importance: small and medium-size businesses too must integrate themselves into the inter-company information and logistics system. These problems apply also for tool and appliance production within bigger companies.

4 General Project Integrated Information System Manufacturing (IIP)

The courses mentioned above are part of a cooperation project between the CIM-Fabrik Hannover GmbH, the University of Hannover and IBM Germany GmbH. This project was initiated in 1986 to integrate CA technology in university doctrine. The students who have thus far received the above mentioned 'CIM certificate' for participation in courses treating CA technology and the corresponding business and information technology subjects are proof of the success of this attempt.

In addition to the educational objective, the IIP project also pursues scientific targets. They include the integration of a company by using a logical centralised database. The scientific targets are also worked on using the model firms of the CFH concept. Within the model 'propeller shaft' company a plan for data integration of the company-specific applications was developed and realised; it will be explained in more detail in the following [Han90].

4.1 The integration model of the IIP project

When implementing a logical centralised database for data integration within an environment of existing heterogeneous applications, it is useful to preserve local data storage of the individual applications. The reasons for this are:

- less modification of existing program code
- no performance losses upon retrieval
- higher autonomy of applications
- support of migration.

The fundamental idea of the IIP approach is based on a so-called integration database (IDB). In addition to local data storage of the applications (CIM components) the database contains complex objects, like production plans or modular system parts lists, as reference objects using a neutral format with system independent attributes. The data or object alignment with the particular data storage for corresponding data modification is controlled by an object handler. In this case it is important that data modification is related to objects. This means that, for example, changes to a work cycle are understood as changes to the object of the work cycle. A locally changed work schedule will be passed on with the appropriate global alteration command to the object handler for propagation to other CIM components involved. This data alignment takes place via a 2-phase commit protocol which guarantees data consistency even in error situations. The object orientation offers considerable advantages for the interfaces needed compared to a link of the individual databases at the record or field level.

In addition to the object handler with its NF2-object-oriented query and manipulation language, the communications component is part of the basic software of such an integrated system. In the IIP approach this component may be accessed by special function calls from individual interfaces.

Emphasis in the IIP project is not only placed on single realisation of a specific system. Moreover, the development provides for a drafting method and the supply of portable software tools, such as the object handler, the communications component and a modelling tool for the IDB, providing portability for various environments. Figure 8 shows concept and elements of the IIP integration model.

Just the design of a specific system requires high flexibility. This is a result of step-by-step realisation and the associated constraints leading to modifications of previously developed modules. The flexibility needed in the IIP approach is

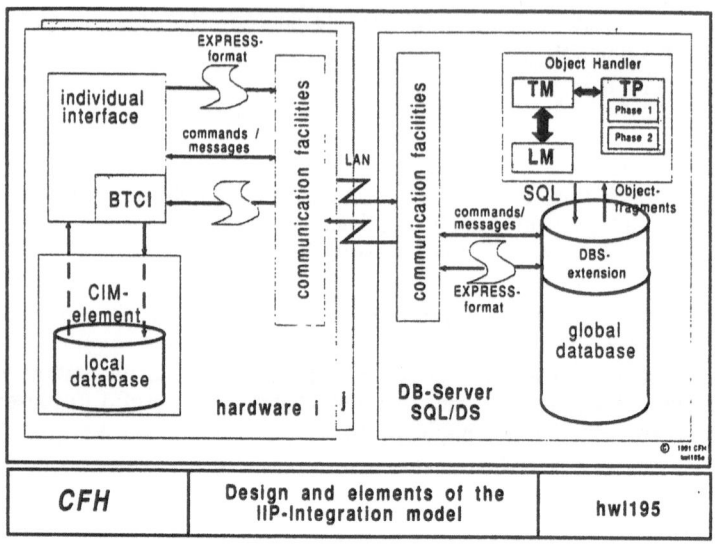

Fig. 8   Concept and elements of the IIP integration model

achieved through the following concepts:

Separation of build time and run time within the object handler

-    Build Time
-    IDB objects are described using STEP/EXPRESS. This information is stored in the IDB for interpreting.
-    Starting from IDB modelling, NF2 operations are generated in executable C-code.
-    The programming of interfaces is supported by a cursor language, allowing one to navigate on the basis of EXPRESS descriptions of objects.
-    Depending on the IDB database system parsers for the recognition of object instances are generated, enabling verification of the structure of received objects.
-    Run Time
-    Generated transaction programs (TPs) access the IDB
-    A lock manager (LM) controls object oriented locks to support even long lasting transactions
-    A transaction manager (TM) administrates competing TPs and controls system-wide and inter-hardware transactions.

Use is made of a neutral exchange format for instances of NF2-objects derived directly from the EXPRESS language.

The widespread networking protocol TCP/IP is used as the basis for the communication components.

To demonstrate this integration concept within the specific environment of the CIM-Fabrik Hannover, a transaction chain is executed during which the existing systems communicate and align their data bases throughout the IDB.

The concept of the IIP project is mainly based on the principles of the CIM-OSA approach for integration which was developed within the EC ESPRIT development program. The correspondence of the IIP concept with the methods and procedures of CIM-OSA will be demonstrated at the opening of the new CIM-Fabrik building [Jor901, Jor902, Kli90].

The integration concept will be used for training and education at the CFH. With this concept and the corresponding individual CIM components it is possible to demonstrate the complex relations and problems which occur while integrating different data processing systems into a comprehensive information system. At the CFH this concept will be further developed so as to integrate the latest knowledge into teaching.

The described integration concept is part of the CFH education attempts. During demonstrations a transaction chain is presented which shows the flow of an order from quotation to order, design and parts planning, to production and reporting back through the integrated system. During this transaction chain all CIM components and the IDB with the object handler are accessed.

5. Summary

With its manifold equipment and integration into CIM technology research projects the CIM-Fabrik Hannover GmbH is able to offer a valuable environment for university teaching as well as for internal company education and training. Within the offered education activities all parts of CIM training may be accessed. Companies may train their employees at the CFH or use the additional CFH capacities to receive support for the realisation of the difficult concepts. This enables the CFH to assist companies in all areas of modern electronic data processing in production area.

Literature:

[Han90] Hanewinckel F., Küspert K., Integration durch objektorientierte Datenbank, VDI-Z 132 (1990), p. 50 ff.

[Jor901] Jorysz H.R., Vernadat F.B., CIM-OSA Part 1: total enterprise modelling and funtion view, Int. J. Computer Integrated Manufactoring, 1990, Vol. 3, NOS 3-4, p. 144-156

[Jor902] Jorysz H.R., Vernadat F.B., CIM-OSA Part 2: information view, Int. J. Computer Integrated Manufacturing, 1990, Vol. 3, NOS 3-4, p. 157-167

[Kli90] Klittich M., CIM-OSA Part 3: integrating infrastructure, Int. J. Computer Integrated Manufacturing, 1990, Vol. 3, NOS 3-4, p. 168-8

142

[Tön91] Tönshoff H.K., Warsch C., Modellfirmen im Zulieferverbund, Datacom 2/91, p. 4-11

[NN90] NN, Kooperation für die Lehre, Hard & Soft, 6 (1989), S. 22

[NN88] NN, Praxisnahe CIM-Realisierung, Industrie Anzeiger, 57/58, 1988, p. 23

Vocational and continuing training in CIM - An analysis of the FESTO/IBM modular training system

Dr. Willi Petersen, Manfred Schön
Institut Technik und Bildung, University Bremen
Bremen Germany

1   Introduction

Approaches to and concepts for computer-integrated manufacturing (CIM) have lost none of their topicality or significance since such ideas arose at the beginning of the 70s, although exaggerated notions of a fully automated "unmanned" CIM factory of the future qualify today as being somewhat outdated. On the contrary, there is, if anything, more debate than ever about questions relating to CIM development strategies and CIM introduction processes, particularly in fields that initially received little attention due to a concentration on "technical" issues. Today, it is above all questions regarding new forms of company organisation and work organisation, as well as changes in the field of personnel management and training that are of great significance in connection with any step-by-step and socially acceptable process of CIM implementation. This can also be seen in studies on CIM implementation which show that approx. 50% of the component effects occur in the fields of organisation (19%) and personnel (30%), and only 23% in technology (Köhl et.al. 1988, p.11). In other words, the introduction and implementation of CIM concepts are no longer to be seen primarily as a, thus far, unsolved technical problem of computer-integrated information and material flow, but as both innovative tasks and the opportunity to obtain a more effective and a more socially and humanly oriented development of work organisation and design. This involves changes in the educational prerequisites and training structures of the employees concerned. In general, more personal skills and greater knowledge of company and organisation are required, the aim being active integration of organisational development; in this way, company-specific experience and complex computer application possibilities in a CIM "Factory of the Future" can be used in a manner that is also open to future potential.

The central focus of this paper thus concerns questions relating to CIM training as part of vocational training, against the background of the inter-relatedness between work, technology, education and training. These questions are related, on the one hand, to the relevant implications for computer-integrated manufacturing and to possible objectives and contents of education. Since objectives and contents cannot be determined purely from educational research and teaching issues relating to work and technology as they are found to exist (see, among others, RAUNER 1991, p. 56f.), our approach here is to try and unravel and substantiate these aims and objectives within the framework of an open curriculum and a prospective education and training concept. On the other hand, we also outline a curricular reference

144

framework for CIM training in vocational and continuing training so that consideration can be given to teachware issues and perspectives developed for an open and transferrable training system concept.

On the basis of the outlined curricular reference framework, and in response to such questions and perspectives, criteria can be developed which allow an initial assessment of the components selected here in terms of their value as learner workstations. The system presented here is the FESTO/IBM Training System, newly developed by the Festo Didactic company in the form of a "Modular Production System" (MPS) and networked with the IBM DAE and AAE "Plant Floor Series" of software. FESTO, the developer and manufacturer of the system, claims amongst other things that the system is "a first step in the direction of CIM training". Besides this claim, it is also of interest to determine the extent to which the system can be integrated as one component in a higher-level integration and training concept and to what extent it contains the prerequisites for this.

It should be remembered that this is neither a study of training practice, nor a comparative analysis of teachware systems in the form of a comparison between the training system and other concepts for a "learning factory" or CIM simulation models, but an attempt at explaining and roughly assessing the educational opportunities provided by these equipment components, taking into account the results of a survey of attitudes towards this specific learning system. By conducting expert seminars with instructors and instructors from the field of vocational and continuing training, the attempt was made to establish a real relation to present-day educational practice by means of the questions asked there regarding existing training equipment components.

2   Discussion guidelines for a curricular reference framework for CIM training concepts

The educational and training issues associated with the research, development and practical fields within "CIM", both technological and economic, are far-reaching and complex, and basically involve any of the very different fields of education from the secondary to the tertiary level. In other words, CIM is not viewed here as a new technology or as a ready-made hardware or software product which only a few specialists are involved in, but as a new "holistic" industrial or company philosophy. This means that managers, information scientists and industrial administrators are confronted by and involved in the development and implementation of CIM concepts to just as great an extent as engineers, skilled workers and stock controllers are. This wholesale involvement of all concerned is based on the very idea of CIM since it contains the strategic notion of fully automated computer-integrated production, from order intake, development and design (CAD), production planning and control (PPC), operations planning and scheduling (Computer Aided Planning - CAP), right up to production itself (Computer Aided Manufacturing - CAM), not to mention quality control (Computer Aided Quality - CAQ) and the final dispatch of products. The logical consequence of CIM, which is conceived of here not only in "centralist" terms, is that its implementation must lead to a reintegration and reorganisation of industrial work (which has so far evidenced a high degree of division of labour), as well as to changing job content in the various fields and departments in any enterprise.

Vocational training, as an independent and significant factor, faces the challenge of producing new and appropriate forms of education and training for this transformation in order to contribute to the development and implementation of both effective and human CIM concepts through exploitation of the scope for active shaping that exists.

If we address the "holistic" aspects of the CIM idea, taking into consideration a curricular reference framework for vocational and continuing training, it can be seen that current changes raise fundamental issues regarding new approaches and the educational objectives and contents which are to be communicated. As is often the case with CIM training concepts, these questions do not immediately and directly relate here to any specific educational, vocational training or specific occupational fields. This is a consequence of the "wholesale involvement" entailed by CIM. The initial basis is provided by existing educational and training concepts which make a relevant contribution to individual aspects of computer-aided office and production work. These conceptual approaches are generally dominated by what are essentially technocentric issues, as is the case with the "New Technologies in Vocational training" framework programme for model and research projects. The outcome of such projects is very often a combination of new technologies and traditional workers' virtues and educational postulates (especially those addressed in the debate as so-called "key qualifications"). An analysis by Matthiesen, referring purely to the field of key qualifications, has shown that 291 (!) different concepts are now in use for expressing the demands that this field of training encompasses (Matthiesen 1988).

The required qualities reads like a kind of "hit parade" headed by the following items (number of mentions in brackets):

-        Thinking in terms of inter-relatedness (17)
-        Ability to communicate (15)
-        Ability to solve problems (112)
-        Independence (112)
-        Ability to work in teams (12)
-        Ability to cooperate with others (11)
-        etc.

These "new" qualifications are an enrichment of previous approaches to and debate on the integration of general and vocational education.

Computer-aided and computer-integrated work, be it office work or skilled industrial work, is now above all a question of the "training and integration" of all employees, and must therefore be related, as far as education and training issues are concerned, to new forms of work organisation in production. Definite forecasts about future training requirements are less and less likely to be the basis on which to proceed, so that the typical adaptation approach in vocational and continuing training is losing any kind of significance it may have had as a result of these structural changes. The crucial issue is therefore how prospective education concepts can be implemented, not only at the planning level within the respective educational fields and occupational structures, taking into consideration the notion of CIM, but also at the level of the specific design of training processes. As far as educational fields and curricular issues are concerned, reference is made here first of all to the five levels within the socio-technical design of organisation structures and computer-aided work systems (see Martin/Ulich/Warnecke 1988).

- 1st level: work structure (socio-technical system)
- 2nd level: division of labour between people
- 3rd level: division of functions between people and machines
- 4th level: man-machine interaction
- 5th level: hardware and software, data structures

Within the transformation process from a Taylorist to a form of work organisation with less division of labour, the specific design of these respective levels in CIM concepts that involve computer-integrated work systems gains central significance. Since the basis assumed here involves open structures capable of further development, fundamental design principles for the reorganisation of industrial division of labour need to be developed which would have to be oriented towards the following superordinate criteria:

- qualified work with highly qualified employees in all fields within the enterprise
- horizontal division of labour for holistic work content
- extensive degree of autonomy

If one takes up concepts for the design of production islands with partly autonomous production groups (see ESPRIT project no. 1217: "Human Centered CIM Systems"), then it is logical in this context that new occupational and work structures will lead in turn to demands for new approaches to the design of CIM-oriented education and training concepts. These must be related in particular to:

- work-related contents of computer-integrated work systems,
- new forms of work organisation,
- the development of organisation.

As far as curricular issues are concerned, the following areas of content provide a basis for discussion with respect to CIM-oriented vocational training:

- organisation structures for the functional fields within industrial enterprises;
- procedures within enterprises (order processing: product manufacture from order intake to dispatch);
- fundamentals of computer, information and communication technology (DP);
- computer-aided work systems;
- historical development of CIM;
- significance, meaning and purpose of CIM concepts;
- various approaches and development strategies for CIM concepts;
- economic and political factors, and how they affect CIM concepts;
- PPS, CAD, CAP, CAM and CAQ as important components of CIM concepts, and their function within industrial production;
- data links between separate CIM components within company LANs;
- topologies, function and utilisation of computer networks;

- recording of personnel, production and machine data (personnel, production and machine data capturing);
- changes to organisation and design of work structures and workplaces through CIM.

Educational fields	**Target groups**
Courses of study Business administration, computer sciences, mechanical engineering, electronic engineering	Business graduates, engineers, computer scientists etc.
Continued and further training Business, computer and technology oriented	Business people, master craftspeople, technicians, skilled workers etc.
Initial training Business and administration, metal-based technologies, electronic engineering	Industrial administrators,4■ skilled workers

Fig. 1   Educational fields and target groups primarily affected by CIM

These fields comprise a curricular reference framework which can serve, on the one hand, as a starting point for making the necessary distinctions between various fields of education and training, between different objectives and between occupational groups (Fig. 1). Such distinctions would have to be made particularly in terms of further curricular developments and new integrated approaches to training. They would have to be related to the more offer-oriented systems for academic, vocational and continuing study and training, and to further training courses which so far have only included particular new fields and technologies, individual systems as CIM components, or basic approaches that combine certain fields, such as CAD-CAM for example. Only in the somewhat unstructured further training field, where there is a greater orientation towards real needs, have there been any more far-reaching courses or training schemes, often designed for a specific manufacturer or factory, for CIM-oriented training. This partly applies as well to new approaches to and concepts for the organisation of learning within the field of continuing training, where attempts are being made with the "learning factory" concept to provide CIM-based training.

On the other hand, the curricular reference framework also provides rough orientations and criteria for the development and evaluation of concepts for learning and for teachware systems, since as a consequence of the inter-relation between the fields of educational decision-making, questions relating to the organisation of learning are interdependently linked to the detail issues with respect to objectives and contents relevant for CIM.

We confine our attention here to vocational and continuing training in the metalwork and electrical engineering fields, and attempt, against the background of the curricular reference framework, to link didactic aspects in the more technically oriented CIM objectives and contents to the question as to how training media are to be developed in future. This is to be achieved, firstly, for questions and perspectives regarding training system concepts, and secondly for the specific example of the FESTO/IBM training system, which is viewed as one component of a broader system.

3   Components of a CIM training system concept for use in vocational and continuing training in metalwork and electrical engineering occupations

If we look at vocational and continuing training for metalwork and electrical engineering occupations and take into consideration in particular those industrial occupations that are concerned with production itself, then it is first and foremost CAM technologies and systems, in addition to the various possibilities for coupling it with other components of CIM, that are of central interest (Fig. 2).

Fig. 2   CAM is the central component of technically oriented CIM training in vocational and continuing training

These involve primarily the new computer-aided skills that have emerged through the reorganisation of the industrial metalwork and electrical engineering occupations, with the consequence that the extended aspects and objectives of computer-integrated skilled work also need to be taken into account. Whereas teaching concepts for computer-aided skills were and are characterised mainly by the separate treatment of particular new technologies such as stored program control systems (SPS), CNC and CAD, we now see that programmable machines and handling systems, flexible production plants or systems for transport and storage, the transformation from computer-aided to computer-integrated work pose a multitude of questions as to how these teaching concepts can and should be revised and developed further in the direction of CIM. The general objectives and content areas in the curricular reference framework outlined above are to be taken into consideration at this point, as are those questions which relate to the specific planning and design of teaching and learning processes at the actual teaching media level, since ultimately it is only at this level that ambitiously formulated objectives can actually be implemented.

If, within the scope of this paper, we restrict the entire complex of differentiated contents to questions concerning teachware concepts and training systems, then the preferred conceptual approach is one which takes advantage of the basic openness and designability of further CIM developments, in the light of the implications for CIM outlined above. As far as present and future computer-integrated work is concerned, an open, modular and adaptable concept will have to be developed. This will have to take into consideration the following design criteria:

-   The equipment should provide integrated vocational and continuing training in computer-integrated work for skilled workers, master craftsmen and technicians in metalwork and electrical engineering occupations;
-   it should contain good methodological options and prerequisites for the development of action and planning competence (manufacturing, assembly, maintenance, diagnosis of breakdowns);
-   it should enable the communication of basic knowledge as well as contents specific for each field and technology, with reference to and with the help of various CIM components (interface technology);
-   to as great an extent as possible, it should be capable of integrating the already available training system components (hardware and software, machine tools, SPSs, etc.);
-   it should be open for the integration of traditional, present and future CIM components (PPS, CAD, CAP, CAM, CAQ);
-   it should contain possibilities for development of the various separate hard and software components (e.g. design of user interfaces);
-   it should enable the use of different network typologies as well as the study of interface (standardisation) problems and system architectures;
-   it should demonstrate a high degree of technical modularity and variability for alternative and future developments (technology experiments);
-   it should be capable of graphically simulating different forms and concepts of work organisation (production in islands, partially autonomous work groups etc.), as well as central and decentral structures (experimental work situations in large and small-scale enterprises);
-   it should permit expandable interim solutions and be able to integrate different planning, development, manufacturing and product components;
-   it should be open for the integration of business and technically-oriented organisation structures and content structures;

These criteria for an adaptable and transferrable training system concept represent highly ambitious demands, conceived of here only as a perspective at first, since no concrete approach for implementing them exists at present. What is important, however, is that this concept opens up an alternative to CIM learning concepts in which CIM exists as a complete, ready-made solution, to be purchased and installed in a specific technical configuration, and for which training material only needs to be developed. Secondly, the openness of the concept, especially with regard to the integration of existing components (not only technical but above all didactic integration), is the intention to show schools and training establishments ways in which they could use their existing equipment in a new concept for skilled computer-integrated work. Possible modifications to such equipment may be necessary, as is also the case when enterprises attempt to integrate various types of "island". The development of training system criteria based on an open and adaptable training system concept, that in practice is also partially institution-specific in design, means that planning and investment can be optimised. Purchases need no longer be made without an underlying concept, as is often the case at present.

The development and planning of an open, adaptable and transferrable training system concept thus involves in particular the design and analysis of possible separate components, which can then be inserted and integrated into a total concept, taking into consideration various educational and technical aspects.

4 Assessments and considerations regarding educational quality and configuration of the FESTO/IBM modular training system

As outlined at the beginning, Festo Didactic has developed a new learning system in the form of a "modular production system" (MPS), networked with IBM's DAE (Distributed Automation Edition) and AAE (Application Automation Edition) software, components of the IBM CIM architecture. This training system for automation technology, through this synthesis of hard and software components, is intended as a means for initiating steps "in the direction of CIM training", whereby the "key to CIM" is to be seen in the DAE software. The system was presented to a wide public for the first time at the 1991 CeBit fair and at the Hanover Fair for industrial goods, and can be viewed as one possible equipment component within CIM-oriented training concepts.

The following is a description of the FESTO/IBM training system and an attempt at an initial media-didactic evaluation and assessment of it. The main question, besides the questions regarding the educational possibilities the system contains, is the extent to which it can be inserted as one component into a more general concept for integration and training system equipment, or the extent to which it has the prerequisites for this. A survey of altogether approx. 30 teachers and instructors from the field of vocational and continuing training was also conducted regarding this training system within the context of workshops that featured the system as a central focus. The assessment that follows is partially based on this survey.

Overview and short description of the FESTO/IBM training system

The training system developed by Festo Didactic must be classified at the uppermost plant level within the Festo MPS concept of levels. This comprises the basic unit, module, workstation and plant levels. At this latter level, several stations each composed of different modules are networked to form a plant. The separate modules are constructed from various basic units used in industry, such as actuators, sensors, mechanical fittings, etc. In this sense the training system is essentially modular, close to actual practice, and open for different kinds of application.

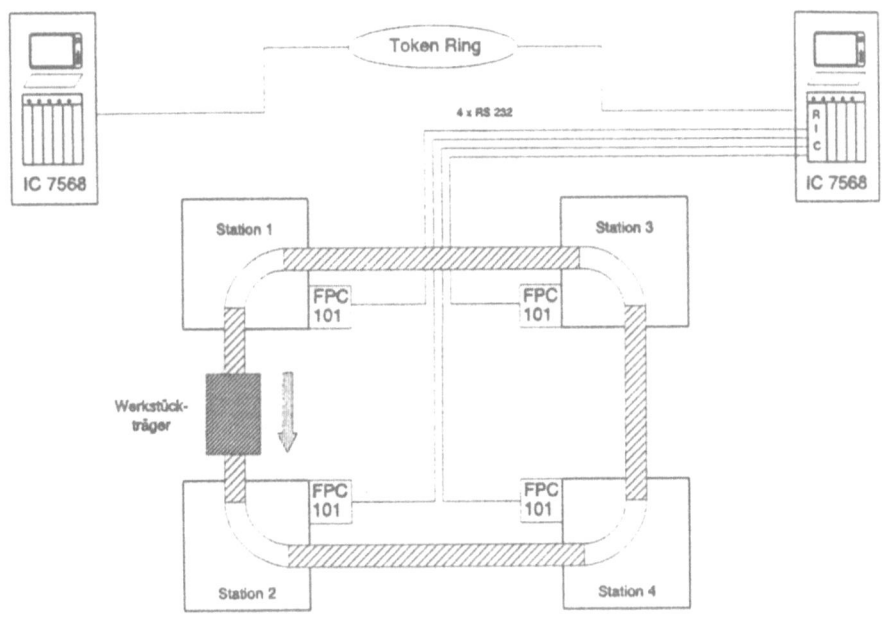

Fig. 3   Schematic diagram of the FESTO/IBM training system

The training system described here was developed by Festo Didactic for a complex plant comprising four different stations (Figure 3). The plant is supposed to represent a flexible production line for toothed gear systems in five distinct variants. The model gear system consists of a mounting plate, one large and one small cogwheel, arranged differently depending on the particular variant. A conveyor belt on which a workpiece holder for this "gear" is mounted provides the mechanical link between the various stations. Each station has its own SPS (FESTO, Model FPC 101) as a controller; this can also be described as a workstation or cell computer. These four workstation computers are networked in a star-shaped network via an RS232 port to an IBMIC7568 mainframe computer used as a host for monitoring and controlling the individual stations. Communication with the linked storageprogrammable control units can be managed with the mainframe computer and a subsystem capable of real time processing. On a second mainframe, also an IBM IC 7568 linked via a Token Ring network to the host

computer, order processing and order placing as well as the statistical evaluation of quality control data can be carried out. Communication and network management functions are carried out by the IBM DAE software, while the process itself can be visually presented by means of the AAE software. The entire system can be controlled manually, in the so-called default mode, or automatically.
Workpiece holder

The following technologies are installed in the individual stations:

Station 1 (Storage and processing technology)

- Storage system for the mounting plates of the model gear mechanism,
- Single-axle feeding unit for the mounting plates,
- Cross-table for positioning the mounting plates for the drilling unit,
- Clamping unit and drilling rig for working the mounting plates,
- Various actuators and sensors, SPS FPC 101.

Station 2 (Storage and assembly technology)

- Storage and transportation system for the knockout spindles of the cogwheels,
- Storage and transportation system for the large cogwheels,
- Three-axle pressing unit for pressing the knockout spindles into the mounting plates,
- Twin-axle assembly unit (handling system) for the large cogwheel,
- Various actuators and sensors, SPS FPC 101.

Station 3 (Storage, assembly and welding technologies)

- Storage and transportation system for the small cogwheels,
- Three-axle assembly unit (handling system) for the small cogwheels,
- Tamping unit for the axles of the two cogwheels,
- Various actuators and sensors, SPS FPC 101.

Station 4 (Quality control)

- Operating tests on the gear mechanism,
- Conversion unit for removing the gear mechanisms from the assembly process,
- Belt and "steering unit" for sorting the gear mechanisms (good and bad parts),
- Various actuators and sensors, SPS FPC 101.

Didactic details and assessment of the FESTO/IBM training system

The main educational information available in written form from Festo Didactic at present is contained in their marketing and sales brochures. This provides only an initial orientation for any didactic categorisation. Roughly summarised, the training system seeks to achieve the following aims:

- the Modular Production System is intended as "the path and a first step towards CIM training";
- the educational fields are stated as "instruction, training and continuing training", without any further specific differentiation regarding occupations or occupational fields;
- the primary teaching objective is "integration";
- the promotion and training of "holistic thinking";
- the promotion and training of "organisational skills";
- the promotion and training of "planning and decision-making competence";
- the promotion and training of "independence in the performance of tasks";
- the communication of "requisite know-how of a technical nature", such as "mechanics, sensor technology, actuator technology and the electronics of systems control";
- the communication of know-how regarding communication through Token Ring networks, process monitoring and process visualisation, network management, as enabled and performed by the DAE and AAE software.
- "learning by doing";
- learning with "industrial components" and confrontation on the part of the learner with "real problems from the field of manufacturing practice".

On the basis of these objectives and other statements contained in the description of the system, certain overlaps of a general nature can now be seen with elements of the outlined curricular reference framework and the CIM teaching concept. These overlaps exist with regard to non-specialised objective-setting and the contents for the CAM component within CIM. However, if this general level is put aside and the training system is viewed under different aspects as a possible component in a total concept, then a considerable number of questions are raised concerning concrete didactic possibilities and limitations to the range of applications, integration or actual operation of the system. In the following, some initial assessments are made in answer to such questions. If we also take into consideration the survey results (stated in brackets), then a broad range of opinions comes to light. The point should be made first, however, that basically it is not possible to achieve the promotion and training of systems thinking or any planning competence solely through hardware or software, but that this is only, and quite decisively determined by the planning and design of the overall methodological concept for teaching.

Educational fields and target groups:

Festo's training system addresses the training and continuing training field only in a global manner. The interesting question therefore concerns the specific appropriateness of the system for different fields of education and different target groups, since only against this background can certain assessments be made or explained. However, given that the aim is a concept for integrated training and continuing training, the fact that this system was deliberately developed for both fields and for a range of possible occupational fields is essentially to be welcomed. As far as the appropriateness of the system for specific fields is concerned, the system as a whole is only suitable for the continuing training field (80%), in the configuration in which it presently exists, and is unsuitable or only marginally suitable for initial occupational training (part-time form) (10%). Such an assessment corresponds of course with a multitude of other evaluation criteria. The training system as a whole is characterised, for example, as being too complex (70%), or the system handling is described as not being satisfactorily solved. A further point that must be taken into consideration is that, because the double teaching hour is the main basis on which learning in the initial school training system is organised, the system is essentially unsuitable for teaching organised in this way. The majority (80%), however, consider it to be well-suited for project-based teaching. A similarly positive assessment is also to be found in the field of full-time vocational training in educational institutions. The possibilities for applying the system are also rated highly for both the metalwork and electrical engineering fields. The system is considered to be best suited, relative to other particular fields, for occupations in the field of production technology.

Assessment of the system configuration

The question as to whether the first step in the direction of CIM training has been taken or can be achieved with the training system, in itself too global a question, was negated rather than affirmed by the survey. The fact that the question was posed in this way shows clearly that the answer to it depends critically on the notions entertained regarding any CIM training and training equipment concept, and on the expectations that exist with regard to any particular system. But the question also concerns the point at which a line has to be drawn, i.e. whether, for example, computer training or training in SPS technique can already be characterised as CIM training. Furthermore, the assessment of the relation between expenditure and didactic utility is important, and must be distinguished from the "theory and practice" of the technical options provided by the system, the extensions to it and the possibilities for integrating it with others, since basically "almost anything" is feasible in the field of CIM today.

We do not fully agree with the assessment obtained in response to the question posed above, because a "step" towards the CIM world was indeed inherent in the approach taken when designing the training system, particularly in its establishing a link between control technology and the IBM software. However, the size of the "step" achieved by the training system - its direction, design, etc. - can certainly be discussed and evaluated in different ways with regard to its educational value. In Fig. 4, which shows the connections between the host computer hardware and the systems and applications platforms, the IBM configuration is placed in its relation to the other CIM components. These are not actually implemented in the training system as such, but are optional components networked within the IBM CIM architecture via a Local Area Network (shown in the system as a Token Ring). To this extent, the training system in its overall configuration is to be seen first and foremost as a subsystem of the CAM component within CIM, accommodating the objectives and contents of skilled work in the CAM field.

The core functions of the IBM DAE software are

-   communication with machine control systems in manufacture (SPS, robot controllers, CNC machines),
-   communication within a LAN
-   connectability to company-wide communication via an APPC interface.

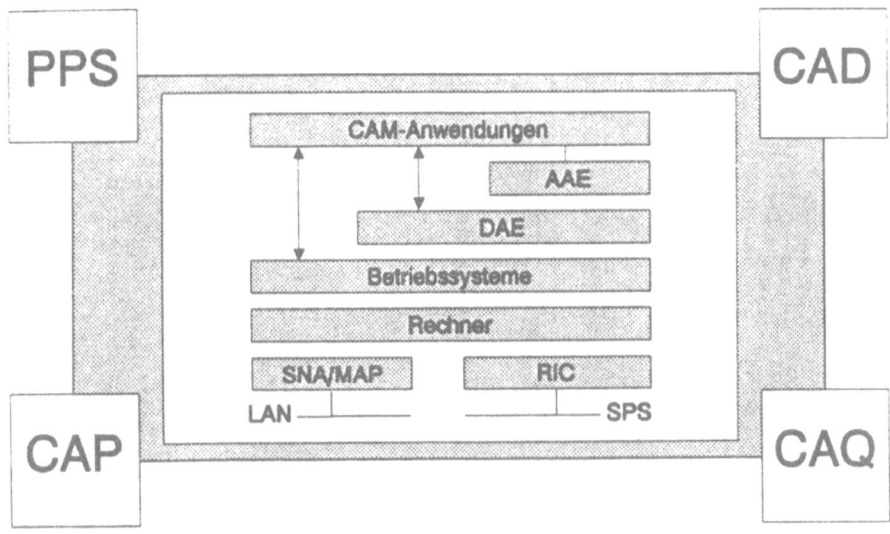

Fig. 4 The IBM CAM configuration within a CIM training system

By means of the options provided by the IBM software, the system is in principle "upwardly" open for networking to a PPS or CAD system, for example. On the other hand, however, RIC (Realtime Interface Co-Processor) boards open up a multi-dimensional and broad communication spectrum in the CAM field at the controller level, so that various machine control systems can be integrated into a network system, as is already the case in the training system for the Festo controller. The AAE software, as developer's tool and execution system, complements the DAE software and provides opportunities for the development and editing of applications for process monitoring or material flow control, using its definition, interface and execution options. The plant and equipment are addressed as logical resources independent of the respective hardware in each case. These possibilities inherent in the IBM software are utilised in the training system in exemplary fashion for:

- Order entry (visualisation of the gear variants and the order table; starting up the variant-specific SPS programs via the IBM host computer),
- Production control and monitoring of the entire plant (visualisation of the four stations and the warehouse),
- production control at station 2 (zoom function; visualisation of the storage and assembly technology) and
- Production statistics (visualisation of quality evaluation for each variant, as well as the number of good and bad parts).

As far as assessing the uses that can be derived form the hardware-software combination is concerned, visualisation of production control implemented in Station 2 is basically to be evaluated positively (approx. 70%), since this opens up a new level at which training can deal with the production control field in a manner

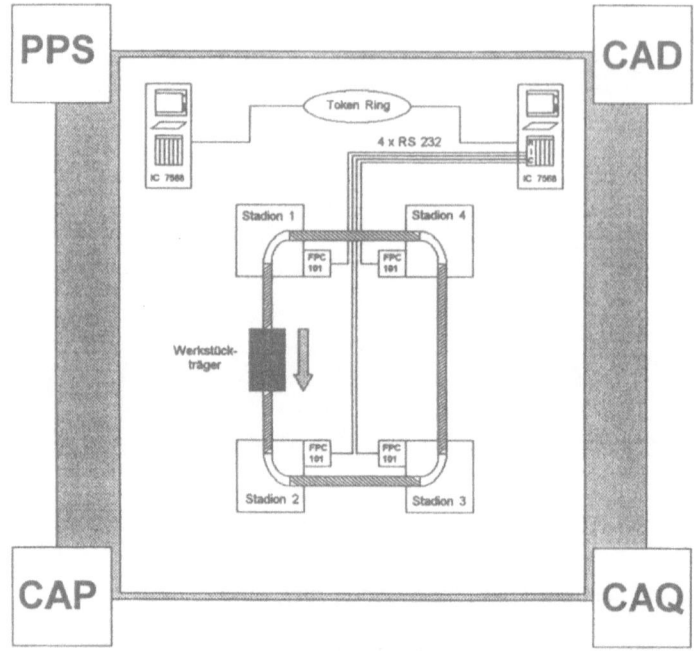

Fig. 5 The FESTO/IBM training system as one component within a CIM concept

akin to actual practice. On the other hand, however, the enormous technical effort and amount of time needed, and the cumbersome handling of the software tools must also be taken into account. These are probably also a reason for the fact that Festo has only visualised one station, or that the evaluation and visualisation of the data available in the host computer from the individual stations via the RIC boards (e.g. on breakdown diagnosis), something that is greatly lacking in the training system (approx. 90%), is largely absent. The existing configuration, which is conceived of for industrial practice and which definitely represents a new dimension in the training field on account of its handling characteristics and the opportunities it presents, is considered too complex and powerful for training purposes (70 - 80%). Within the framework of a total concept, the questions which would have to be clarified and researched would be whether the existing openness and scope for development with respect to industrial applications can also be used in the training field in the form of "interim solutions" yet to be developed.

If we now look at how the training system was assessed with regard to the direct communication between the host computer and the individual stations, then firstly it can be seen that the control technology implemented in the plant as a whole enables a broad spectrum of objectives and contents in this field to be taught. By means of the sensors and actuators in the stations, approx. 120 inputs and 80 output signals can be processed and controlled vis the SPSs. Of these input signals, about 50 go to the host computer and could theoretically be processed and utilised there, in the same way as the training system does for a small part of the training system, as described above. Further processing of the information, e.g. processing of sensor signals from the stock surveillance in the stations, could be done by and for a Production Planning and Control system. Communication with the equipment is very limited, however. Like the star structure of the network, it is greatly dependent on the Festo FPC 101 SPS that is used. Since this SPS model cannot work with a bus typology and does not support the MMS functionality (Manufacturing Message Specification, ISO 9506) necessary for further development, this means, for example, that the network topology cannot be changed (e.g. to a bus structure), that no data exchange between the various stations is possible and also that the communication services that are available are very restricted on the whole. Further development and opening of the system, also for SPSs from other companies, is required to facilitate the integration of other automation and existing equipment components. This would also make it easier to extend the modularity of the system as is proposed here. This would then facilitate the sub-division of the plant into separate stations and also enable, for example, the relatively autonomous treatment and linkage of control technology in only 2 stations.

Summary assessments

The description and assessment of the training system included at the same time the direct or indirect presentation of CIM-oriented objectives and contents which can mainly be assigned to the CAM component within CIM. They can contribute in particular to the extension and promotion of skills and competences in the field of control and information technology. Provided that appropriate methodological concepts and organisational preconditions for learning exist, the training system is also conducive to other objectives of a more non-specialist nature, but which are of increasing importance for skilled and qualified computer-integrated work. The training system is amply suited to the recognition of causal interconnections, for example, the training of system thinking (approx. 80-90%), and to the promotion of holistic thinking and team work (approx. 70%).

Through the integration of components from industrial practice (hardware and software) that are on the whole genuinely work-related and technology-related within the FESTO/IBM training system, contents and processes can be communicated and taught using the system. It thus provides a good basis (90%) for the integration of theory and practice that is being increasingly sought after in training concepts today. As far as the question of an open and variable design of different technological or organisational concepts is concerned, however, the possibilities of the system are to be assessed as being rather limited. With regard to the modularity mentioned in the description of the system, for example, it must be noted that the system in itself indeed possesses a high degree of mechanical modularity, but that, overall, it appears somewhat "compact, closed and of low accessibility" in its existing configuration. The preconditions for manipulating or expanding the system, or linking it in a didactic-methodological way to existing training equipment components, are thus considered to be somewhat lacking (approx. 70-80%).

In summary, it should be mentioned with respect to the survey results that in the workshop seminars conducted with the training system, the descriptions and assessments presented here were discussed in conjunction with many other detailed questions, all of which could not be dealt with here. Discussion of these questions was sometimes very controversial. All in all, however, given that new territory in the field of teaching media was opened up with the development of the training system in this form and that further developments in this direction are necessary, general trends are also recognisable which, taken together, provide important pointers for the current debate on CIM-oriented training and equipment concepts.

Literature:

Köhl, Eva/ESSER, Udo/KEMMNER, Andreas/WENDERING, Andree: Auswertung der CIM - Expertenbefragung (Februar 1988). Aachen: Forschungsinstitut für Rationalisierung an der RWTH Aachen, 1988.

MARTIN, T./ULRICH, E./WARNECKE, H.-J.: Angemessene Automation für flexible Fertigung. In: Werkstattechnik, (1988) 78, S. 17 - 23.

RAUNER, Felix: Soziale Gestaltung von Arbeit und Technik - Perspektiven einer Leitidee für die berufliche Bildung. In: HEIDEGGER, Geradld/JACOBS, Jens/MARTIN, Wolf/MIZDALSKI, Reiner/RAUNER, Felix: Berufsbilder 2000. Soziale Gestaltung von Arbeit, Technik und Bildung. Opladen: Westdeutscher Verlag, 1991 (Mensch und Technik. Sozialverträgliche Technikgestaltung; Band 18. Hrsg.: Der Minister für Arbeit, Gesundheit und Soziales de sLandes Nordrhein-Westfalen), S. 41 - 92.

CIM Training by means of simulation models

Dr. Herbert Tilch
Institut Technik und Bildung, University Bremen
Bremen Germany

Abstract

CIM training by means of simulation models is not only a instructional concept but based on development and implementation of simulation as a part of an in-plant planning and production process. Simulation models have been set up for different dynamic processes from singular technical aspects up to the most complex production systems. For CIM training the more complex and dynamic simulation models are of most interest. The question of shaping and simulation of the production process is a focal point of attention for prospective CIM training.

1 Different types of approaches to the simulation system

There are different approaches to setting up a wide range of implemented simulation systems and the corresponding CIM training. These approaches vary mainly in their social aspect.

1. Approaches to open structured simulation systems. Technical and organisational aspects of the corresponding production systems are included but reduced to a given structure without the communication and social interactive process. The "Flexible Production Automation Simulation System" (Alshahid/Thomas) is an example of a developed complex system in this kind.

2. Approaches to maintaining the social and decision-making process. Simulation in this case has to be understood as part of a reinforcement of the quality of decision-making within the scope of the considered variables or alternative scenarios such as variation of perspectives, consequences of assumed decisions and concepts of participation. "Simulation in Work and Technology" (Scheel/Bruns) is an approach that emphasises this aspect of simulation.

3. Approaches with a focus on the social aspects of the working process. These approaches are based on the idea that human and organisational factors are the most important factors of success in prospective CIM concepts. To innovate the work process means to increase the capabilities of all those involved in the decision-making process. The contribution of "Simulation and Critical Factors" (Bolk) contains such a model with an emphasis on the training aspect.
Simulation as a development in problem-solving capabilities does not depend on the use of computers but on the social environment of the work system situation.

All the different approaches have the double functional aspect of simulation: On the one hand, simulation is to represent real processes of a special kind and to show alternatives or partial alternatives that can be discussed for other solutions. This part aims at the shaping process of real dynamic situations. On the other hand, along with this object-oriented function simulation has a teaching function. This function will become more and more important, particularly in strategic and social simulation systems. Simulation having this function trains people managing and operating in complex work systems.

According to the learning function, two aspects have to be considered:

1. Learning by simulation means learning in a situational context, where the context is only a media for another complex situation. Computer-based simulation is attended by a learning process in which cognitive structures are developed while oriented to a simulated situation represented by the simulation structure. Simulation aims to provide an experience through the simulation system.

2. Learning by simulation involves a transfer problem. Successful operations carried out by a trained person in a simulation environment are not equal to reasonable behaviour within the framework of particular situational factors in the production process. There is a difference between the simulation system and the real system and corresponding learning and operating attitudes.

2 Application of simulation systems

The main types of different application fields will be considered only with respect to technical aspects of the systems. In a more comprehensive consideration three levels of application can be described as follows:

■ Simulation in operational technical fields. Simulation system is to be present a realistic interface and an easy-to-use display. Many simulation systems support control systems in production technologies, such as operation of CNC machines (turning, milling, laser cutting, etc.) or in particular operations, such as collision check. Simulation systems help to control the movement of robots in industrial applications.

■ Simulation in planning systems. There are advantages in using planning simulation systems by analysing different variables and solutions and by developing strategies and optimising solutions. Fields of application are

- production planning and control systems
- work planning and logistic systems
- configuration systems by integration of modules
- production system and planning systems oriented to CIM

Regarding the use of simulation systems in the field of production planning and control systems (PPC):
Most of the systems have integrated parts of simulation systems for time and/or capacity study. With this instrument a variety of coordinations of orders can be

simulated via a given structure of equipment and facilities. The consequences according to capacity utilisation and production coordination can be simulated and used for a optimised decision.

■   Simulation of a social process. In this field you can assign a wide range of planning games for economic and business management. However, the main part of this field of simulation involves the training process for employees in finding solutions in a simulated environment.

Traditional simulation systems can be adapted to a classified application field, but the adaptability is greatly limited. The special feature of a given arrangement or a structure of a situation to be managed in practice is much more complicated and involved in the total management process than a classified task can describe.

An important step in developing simulation systems is automated model optimisation. The reproduction of situations and processes in a model is only a small support for the user. By using a learning simulation system, the system reduces time and adapts itself to the input variables.

A highly sophisticated system is, however, only an instrument for a qualified user. In search of operational strategies the creativity of the user is required. There is an interdependence between the system's ability and the user's ability. For a creative user the simulation system is a helpful tool. The responsible and creative use of implementation and application of a system requires a user with relevant training. It is the user's task to choose the appropriate solution, to compare a possible situation with his own experience and to decide about continuing the search for an optimised solution or not.

A further step in simulation system development is in the area of system flexibility. As shown in the papers of Thomas/Alshahid and Bruns/Scheel, the systems become more and more complex in order to provide more flexibility in the real production or planning situation. Such a system can be described as follows:

1.   A flexible simulation system is not based on a simple algorithm to represent the real situation in a model. It is intended to integrate simulation systems into the company's organisation.
2.   Training and participation are important factors in software engineering concepts for development and implementation of flexible simulation systems. This is, for the most part, equivalent to the general path of creating open and flexible production systems.
3.   Modularisation and open structures in system application and system use.

3   Benefits and limitations in using simulation systems for training

Generally speaking, several advantages can be assumed by integrating real production processes into simulation systems. For instance:

- Simulation can be operated without any danger for real life systems. This is important for training in difficult and dangerous situations of a process state.
- Simulation reduces the time element in time-consuming processes. Results of motion studies of a process can be achieved in different states or different inputs in a very short time.

However, the benefit of simulation in training has been limited due to differences in the system's environment. People will not change their attitudes to the real situation but to their assumption and reflection about the situation. Training in simulated situation can be evaluated but there is only little success in predicting attitudes in the real life situation and regarding the background knowledge to assess the comprehension and the cognition process.

In using simulation systems for a production process, realistic training aims at improving decision-making attitudes using auxiliary knowledge associated with such systems.
Therefore, it is most important that

- the simulation system is appropriate to the complex production process, resembling that part of the production process from which the simulation system has been taken
- the training participants are provided with background knowledge to assess the process and for efficient communication about the solution for integration into the real life process.

Application of technically oriented simulation systems in the organisational process is biased in a sense. These systems normally describe this process from a point of view that is appropriate to a full automation process. As for the application in a process that can be described by managing the equipment the simulation system might be represented by a algorithm resembling the rough description of the process. The role of the worker is reduced to an operator in an ongoing technical process. The creativity to manage this process has been shifted to the system planner and programmer.

The representation of a complete production process should, moreover, support "social simulation" and shaping of the working system. Human communication and the social process should be included in a comprehensive manner of training via CIM simulation. Appropriate methods for training in this field can be found in situational learning of creativity, cooperative self-organised learning, and managing social and communication parts of a production process in a working team rather than through technical training. There is no simulation system that includes all technical and social aspects of the production process with more than a simple heuristic function. Simulation systems can be used as a tool for training management.

Regarding the social abilities and communication aspects, social situation simulation, as reviewed by H. Bolk, can be combined with technical systems. The development of coordinated systems of all aspects of the production process will be an important task for the near future in CIM training. For a prospective model it should be mentioned that the participants' qualifications are an important part of the training system. The auxiliary facility of the production process should be embedded in communicative process control. Technical simulation systems should be used and adapted in more or less self-organising teamwork. Efficient and human-centered work requires computer-based simulation and social interaction training.

FPASIM - A System Approach to Human-Centered, Flexible Production Automation (FPA)

Dr. Kuteiba Alshahid; Roland Thomas
ERT euro-technik (UK) Ltd, Brigton
Brigton England

Abstract

Many attempts have been made in the past to automate selected functions or islands within industrial plants. However, the benefit has been limited due to the lack of efficient communication and integration between these automated functions and/or islands. At a later stage full automation of some factories has emerged. Nevertheless, little success has been reported in a small number of these projects in Japan and some were a total failure, as at General Motors in the USA and at the 'Halle 54' project at Volkswagen. This is because most of these systems ignored the role of human effort in conjunction with computers.

In contrast, FPASIM, which stands for Flexible Production Automation Simulation system, stresses the flexibility of the automation processes by recognising that a human-centered approach should be used in conjunction with computers for the automation of any process. It appears, for instance, that flexibility is not so much determined by technology as by people assisted by computerisation and automation. The flexibility of the organisation and the quality of production are determined by communication, whether or not formalised, between people and between people and machines [1].

1   INTRODUCTION

The FPASIM system is a single integrated modular computer system for the simulation and management and control of Computer-integrated Manufacture, developed for use in small to medium-size industrial factories and in technical education and learning environments where it can be used as a teaching tool and learning environment for both students and supervisors.

The system is aimed at serving different types of users with different functions and responsibilities. However, special attention has been given to the lower echelons of the work force rather than senior management. Thus, middle management and supervisory functions are the central focus of the FPASIM philosophy. These are the links between management and production which include personnel dealing with planning, preparing and controlling production and/or assembly. To impart a sense of realism, all company functions and their interactions are simulated.

2   FPASIM COMPONENT STRUCTURE

Whatever method has been adopted to provide intelligent simulators resembling in some cases very complex real life automation processes, it is most important to understand the building blocks and the auxiliary knowledge associated with such systems. This is to assess the comprehension and the cognition process for such complicated systems. Thus the complete FPASIM system has been divided into four interconnected components: macro-simulator, micro-simulators, ISC (Integrated Shop floor Control) system and finally the courseware systems.

The courseware systems are several independent computer programs identifying essential aspects encountered in most of the automation processes. They provide elementary skills and comprehension by means of knowledge description, measurements, identifications, classifications and illustrations.

The micro-simulators are specific detailed simulators relating to several important functions within the automation process.
They provide the application environments and the detailed analysis for the knowledge learned using the courseware system.

The macro-simulator is the main FPA simulator and is the system manager and control unit. It combines all the previous elements of knowledge to provide the synthesis and evaluation needed to design, reconstruct, judge and justify its actions.

Finally the ISC system, which is the main DNC system, provides links to the previous components and industrial shop floors.
The above four components are all connected via a standard Local Area Network so that data can be shared and actions conveyed to the rest of the components when necessary.

Each of the above components can also be used as a stand-alone system to serve a particular aspect of the automation process. We will describe each component and show the relationship between them.

2.1   THE MACRO-SIMULATOR

The macro-simulator can be described as a multi-function integrator and a control and management simulator for a wide variety of production processes. It is a reprogrammable distributed processing system concerned mainly with the organisation of information. Its objectives are to provide learning on how to function within the complexity of company functions, which by means of automation must result in a flexible production process. To this end a scale model is used where all major functions of industrial factories and their interactions are simulated. However, this is based on partial
automation allowing the simulation of real world industrial problems and the necessary communication between all related functions. In this way we avoid the impression of working on an ideal laboratory model and practice on the basis of all possible solutions and decisions in such cases.

2.2   THE MICRO-SIMULATORS

These are several interconnected graphics-based system models which provide computer solutions to education and industry for selected functions of Computer-integrated Manufacturing. These systems are an invaluable feature of the FPASIM system developed specifically to teach and train the user of all aspects of a production process. Since it is very expensive and very time-consuming to have access to the actual hardware devices, the micro-simulators by contrast offer the user of any level and at any time the possibility to experiment and learn much about the production process. Thus, robotics, CNC machine tools, writing CNC programs, etc. can all be understood even before using the macro-simulator of the FPASIM system. The output produced by the micro-simulators models can easily be transferred to other component of the FPASIM system. We should note here that the area of CAD/CAM is covered in the micro-simulators component.(c) ERT (UK) LTD 9/91

2.3   THE ISC SYSTEM

The ISC (Integrated Shop Floor Control) system consists of four interconnected graphics-based industrial systems used in conjunction with the ERT industrial controllers to provide complete automation between a PC server and any machine tool, thus supporting CNC file management, downloading CNC component programs or other programs from a server to a machine tool and vice-versa. Also provided is real time machine tool monitoring and statistical data analysis, tracking of tools, fixtures and jobs, real time machine tool sensor monitoring and periodical data analysis and finally order planning and cost. These systems are applicable for one machine, a group of machines or for the entire shop floor. Figure (1) shows an overview of the ISC system.

As Figure (1) shows, the ISC system is used with the ERT industrial controllers (e.g. IC101) which are intelligent industrial controllers providing the communication interface to transmit and collect data (e.g. CNC programs) via the network to and from a machine tool. They can also monitor the status and performance of these machine tools both manually and automatically for input and/or output.

2.4   THE COURSEWARE

These are fully interactive sets of independent graphics-based systems, each of which deals with a specific subject (e.g. reading technical drawings, CNC lathe cutting, welding, etc). They include administration systems which provide facilities for generating reports. The courseware can be used by students as a learning process and by teachers as an open learning teaching aid.

2.5   THE EDUCATIONAL STRATEGY OF FPASIM

As can be seen, the above classification of the FPASIM component structure is in fact allied to the level of behavioural and learning objectives within the cognitive domain from mere knowledge of facts to the intellectual process of evaluation. The courseware, for example, results in the student being able to identify and describe all the fundamental knowledge which lays the foundation for the succeeding elements in the learning process.

The micro-simulators take the student to an intermediate level in the learning process by introducing application environments so that manipulation and analysis of different situations are performed. They provide a focal point for extending his/her knowledge through constructive discussions.

The macro-simulator, contributes to an advanced level of understanding in the learning process. Thus, by evaluation and synthesis, the student learns the impact and the effect of one action in relation to the rest of the functions within real life simulated environments. He/she can then be critical of any decision made which opens the way to making judgment and justification relating to value criteria. [6]

3 FPASIM SYSTEM ARCHITECTURE

Several types of computer machines and operating systems and environments are used for the entire FPASIM system. The first component, i.e. the macro-simulator, is located in the most powerful machine. An 80486 microprocessor based workstation is provided as the computer server for the entire FPASIM system. This runs the UNIX operating system since the latter supports multi-user and multi-tasking configuration. At the core of this machine server lies a Relational Database Management System. Several Rule Base Systems (i.e. Expert Systems) are set up around this database.

The X-Window (network transparent graphics system) has been chosen for this machine using the Motif toolkit. Up to 16 X-terminals can be connected to this workstation at any one time.

The X-window systems and the number of terminals connected to the macro-simulator (an amount of 16 to 20 MB of RAM) are needed to cover all possible cases of the simulation.

A number of less powerful computers running the MS-DOS operating system are used for the rest of the components of the FPASIM system. It must be noted that more than one component could be situated in one machine. However, we divide the major three components at least into three different machines namely the ISC system, the micro-simulators and the Courseware. The human interface for these three components is supported using several graphics environments, such as the Graphics Environment Manager GEM or Microsoft Windows (3) or the ERT Teachers Tool Kit (TTK). These three components are also offered with the UNIX operating system. In this case their graphics interface is the X-Window graphics system. All four components are interlinked via a standard Local Area Network or a combination of standard networks which make use of the ISO/OSI communication standard which is based on the MMS protocols. Note that any MS-DOS machine could also be used as a X-terminal for the macro-simulator.

4   FPASIM SYSTEM DESCRIPTIONS

Having identified the four major components of the FPASIM system, we will now restrict our descriptions to the macro-simulator, which is the component that provides the FPA environment simulation. The macro-simulator at its highest level consists of two phases: a simulation phase and a management and control phase.

4.1   FPASIM (SIMULATION PHASE):

To date, simulation technology has not succeeded in bringing about good examples representing real and complex industrial situations. This is due to many reasons, some of which were discussed earlier in section (1), the most important of which is the assumption that simulation should be done in ideal environments.
This often leads to a deceptive understanding of the nature of work in real life situations and ends up in providing superficial and incorrect decisions and solutions.

The important issue undertaken by ERT is the realisation that simulation systems should not be designed to operate in these ideal environments. Real life problems are far from being ideal and for any simulation system the rate of success and failure will depend very much on how far any system can achieve a realistic solution for a real life complex situation. Thus our attention has been directed at providing these issues and the necessary means so that the combination of complexity and flexibility can be achieved and transferred to a simulation environment. An outline of our major points:

1.   First of all we do not attempt to provide systems that operate in an ideal world.

2.   We must understand several issues, such as the (socio) organisational integrations and communication for any automation system. The established order or culture of a company plays a major role in the implementation of any automation system ([1],[7]).

3.   There is a growing awareness that personnel training is an essential factor in the implementation and deployment of FPA systems.

4.   Training on any simulation system should be modular. A module is an amount of subject matter to be completed as a unit. It is important for the scale model to grow along with the developments in industry. This means that the scale model must have a modular structure so as to relatively easily accept innovation in specific areas ([1], [7]).

5.   It is necessary for trainees to learn how to function in a system, not to be involved in learning the technological aspects of such systems. The technological interpretation and implementation of a flexible production automation system greatly depends on the product situation and is not necessarily rule-determined.

On the other hand, and as far as the system architecture is concerned, many simulators rely on conventional data processing methods for the performance and interactions of these systems.

We believe this is a major shortcoming. Data is a deterministic representation of facts and real world information is incomplete. Total reliance on such systems has led to unrealistic performance and is time-consuming for both machines and manpower. To overcome this problem, we have chosen a combination of relational database systems and knowledge-based systems (i.e. expert systems) so that corrective actions under certain conditions can be taken simultaneously. Thus we have to provide the ability to optimise and incorporate human knowledge and experience. The FPASIM system can simultaneously perform educated guesses when necessary to recognise the right approach to problems and to provide efficient and realistic suggestions when data is incomplete.

The simulation phase of FPASIM contains two independent modes:
Single and multi-user mode. However, the multi-user mode can also be used as a single user mode. In both modes the system here is an interactive graphic button and menu-driven information processing system.

4.1.1   SINGLE-USER MODE

This mode is used by a single user on one machine only (via the X-terminal or via the console) which allows the experimentation of the design and the simulation for a wide variety of automation processes. It provides comprehensive simulation for a variety of flexible manufacturing cells, assembly cells based upon the JIT philosophy, automated warehouses, master production scheduler, quality Control, etc. For example, if an FMS simulator is required for a certain cell, the system then starting with the design of the intended cell, prompts the user for the number of devices available, the type of devices, such as the number and the type of any CNC machine tools, robots, inspection cells and any additional information needed. Once the design is completed, simulation may take place. Special 'What If' facilities are incorporated so that the simulation can be stopped at any time and the user can interrogate the system for any possible change of the configuration or the setup of the designed cell. The simulator in this mode produces output describing the best possible configuration and the order of execution for the available hardware devices. This output is made available to the management and control phase of the macro-simulator (see later in section 5.2).

4.1.2   MULTI-USER MODE

In this mode of the simulation phase several users on several X-terminals can interactively engage in simulation tasks where it is possible to accommodate up to 16 users at one time. Every task is executed as a game and every game is played by a number of users with a supervisor or game manager. These games are in fact a training workshop/factory which could be described as a company where users occupy various jobs in a job rotation system. This is to allow every user to experiment in detail all the relevant functions within the production process.

Each user is allocated a terminal. Each terminal represents a different function within the automated factory and all terminals are connected to one machine so that data and communication are ensured between all the users.

The game is initiated by the game manager, and for every game there are several cases (files) which can be played. Each case stresses a certain aspect of an automated process. The interaction between users is monitored by the game manager and is conveyed to the rest of the users. Figure (2) shows an overview of the multi-user mode of the simulation phase.

4.1.2.1   EXAMPLE:

Let us assume that a certain automation process has to be simulated. We want to describe to all users all the major functions involved in this process, the role of every function and its effect on other functions, the possible integration and communications needed between these functions and their operators and the order of execution in a real life environment.

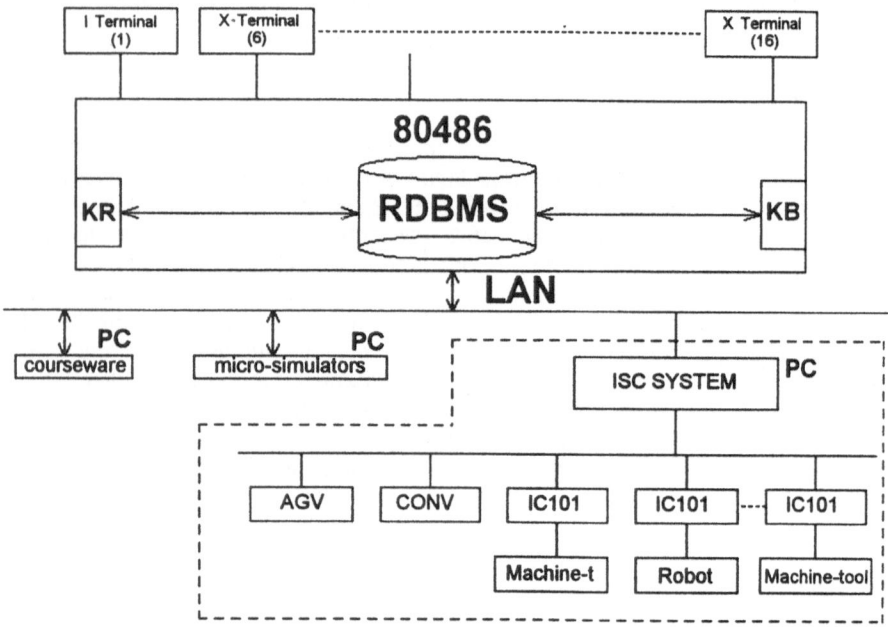

Fig. 1   FPASIM System Architecture

The game manager first of all initiates this game to suit the number of the available users (for example 8 users). Then each user is allocated a terminal. Each terminal performs a different function of the automation process. The user then becomes the named function in this automated process. As indicated earlier, we should note that these games should be played on a rotational basis.

Thus terminal one can be assigned to the 'Design' section, terminal two to the 'Process Planning' section and so on up to terminal 8 as the 'Quality Assurance' section. The game manager can communicate with any user at any time and has the capacity of interrupting any user while users can communicate only with each other. To bring the simulation process into the realistic world, the game manager may start the execution of the game from any section, e.g. terminal 6 which is the 'production' section. Thus starting from Terminal 6, the user is asked to manufacture a number of jobs. This is displayed on his/her screen terminal as a number of icons representing the goods to be manufactured. By clicking on the icon representing one of these jobs, more information is displayed representing the current status of that job. The user then has to check with other players (e.g. the warehouse section) for the availability of raw materials and, based on the reply, he/she may communicate further with other players to provide the target job. Meanwhile the game manager has already given other instructions to the other sections with different tasks. We should stress here that all these activities are carried out parallel to each other while the game manager is monitoring all the players. The game manager can intervene and even stop the interaction of any user for further consultation with the game players.

4.2   FPASIM (MANAGEMENT AND CONTROL PHASE)

The management and control phase of FPASIM which can be used either as a stand alone system or in conjunction with the simulation phase is responsible for the actual management and control of all hardware devices within the chosen production process. It offers a complete FMS control configurable unit. This includes any number of CNC machines, robots, conveyors, AGVs, assembly lines, industrial controllers, PC's etc. Once the system is configured, then this part of FPASIM controls and supervises the execution of any FMC. While the control is in progress, an animated picture accompanies the execution of the program step by step.

5   CONCLUSION

FPASIM is a single integrated distributed processing system for flexible production automation concerned mainly with the organisation of information. It is based on partial automation by recognising that true flexibility for automation processes can only be achieved through a combination of humans and machines.

It consists of four major components, each of which serves a particular aspect of the automation process. All major functions and their related knowledge are simulated. It is designed to simulate real world situations not the ideal situation. Integration and communication between all its components are of special significance. Industrial standards and modularity are the central focus of its design and behaviour.

6   ACKNOWLEDGMENTS

The authors would like to thank many of their colleagues at ERT (UK) and The Wihold Group (NL) for their invaluable suggestions and comments about some parts of this work. Peter de Jong of 'Stichting Opleidingen Metaal' (SOM) of the Netherlands provided invaluable advice about the multi-user mode of the simulation strategy.

REFERENCES

[1]   FPA TRAINING: THE SOM VISION, P. de Jong, Stichting Opleidingen Metaal, February 1991, Netherlands.

[2]   FPA/MTO PROJECT, Presentation of CIMSIM for FPA/MTO project, Almere-Stad, ERT .euro-technik, February 1989, Kuteiba Alshahid and Roland Thomas, UK.

[3]   Modern Manufacturing Systems Design & Control Principles, Dr. John Parnaby, A joint training project, Lucas Group .Training,Manufacturing Systems Engineering, Lucas, UK.

[4]   Manufacturing Database Management and Knowledge Based Expert Systems, Paul G. Ranky, 1990, CIMware Limited, UK.

[5]   Flexible Manufacturing Cells and Systems in CIM, Paul G. Ranky, 1990, CIMware Limited, UK.

[6]   Teaching in Further Education, An outline of Principles and Practice, L. B. Curzon, Second Edition, Cassel 1983, London, UK.

[7]   Developing European Learning Through Technological Advance, DELTA - Exploratory Action, Final Technical Report, Delta, May 1991.

Terms Used

AGV	(Automated Guided Vehicle)
DNC	(Distributed Numerical Control)
FMS	(Flexible Manufacturing System)
FMC	(Flexible Manufacturing Cell)
IC101	(The ERT Industrial Controller)
ISC	(Integrated Shop floor Control)
ISO	(International Standard Organisation)
MMS	(Manufacturing Message Standard)
OSI	(Open System Interconnect)
TTK	(Teachers Tool Kit)

Simulation and Critical Human Factors; the Dutch Contribution to CIM Simulation

Dr. Henk Bolk
InterVisie
Leiden Holland

1 Introduction

The concept of "critical success factors of CIM" can be approached in a variety of ways. Depending on the approach to CIM and the expectations resulting from it, one will have different expectations about the tools to support the development and training process.

Let me first mention some examples of approaches to CIM and the resulting expectations. An outline of the critical success factors for CIM and its subset of critical human factors for the implementation of CIM will be followed by the support to be expected from simulation. The paper will finally deal with the Dutch contribution to CIM simulation.

Approaches to CIM:

- "They tell me that CIM is a relevant technological innovation; Now you tell me whether it is relevant for my company or not"
- "CIM is a means to increase profitability of my company; tell me what success I can expect"
- "We no longer can ignore CIM; tell me how to successfully use CIM"
- "I want to proceed with CIM applications; tell me how to successfully cope with the accompanying changes in my company."

Four different ways of looking at CIM. According to these four attitudes, the degree to which CIM is defined as a way of fundamentally influencing the company increases.

I cannot predict the attitude of those present. A German report, however, describes in what way the CIM strategies of companies can be characterised, and your expectations are no doubt linked to the characterisation that best fits your situation.

Category I

In the first place, there are companies that gradually use more advanced technology by steadily investing in the replacement of old equipment. They deal with vendors of equipment that confront them with the increasing possibilities of hardware and software. In a number of cases in these companies the term of 'flexible manufacturing' (or even 'CIM') has become a receptacle for the collection of available advanced production means. Like someone who buys a BMW with advanced engineering technology and electronic driver comfort simply because it is available at a reasonable price. There is no mention of an intended company or change strategy in the field of automation, let alone a strategy that takes into account technical, organisational and informational components. This does not seem to be necessary because the company does not intend to fundamentally turn around organisational practise.

Most companies in the Community belong to this group. A typical comment which was made by one of our customers: "Can one still buy machines that are not CNC-controlled?"

Category II

In the second place there are companies in which one realises that, because of current technological developments, more structural changes will gradually have to (or in fact) take place in the organisation of production and in the training of employees. The use of production technology leads to doubts about whether available options are utilised optimally. Through literature and (especially) through discussions with colleagues and vendors one learns that organisational changes may allow for a more effective reaction to changing training needs, to the mutual coordination needed, to delivery, quality and stock difficulties, etc.

These companies have been given the term "structure-seeking". Their orientation is especially 'reactive'. The anxiety about the structural changes needed stems from apparent malfunctions in production.

Category III

In the third place there are companies that feel that optimum use of available production technology also means structural changes of the organisation. The strategic framework of the company deliberately takes into account the demands about time-to-market, quality, faster changing and more strict customer needs. One intently wishes to control and manage the integrated change process of technology, organisation and information.

The available technology has a degree of advancedness that requires optimum use in a flexible organisation with completely new structures: new and dynamic structures of production, people, information, decision-making, consultation, cooperation, control, planning.

These companies are described as "structure-innovating". They are proactive. The need for changes in the company structure stems from the conviction that production technology and organisational factors need to be integrated in a dynamic interplay.

The remainder of this paper will elaborate on four issues:

-   The critical factors for CIM in general, by elaborating on the factors that play a role in the approach to CIM as it is practised by companies in categories II and III
-   The subset thereof regarding the critical human factors; in addition to that, we shall formulate some essential prerequisites for accountable and sensible tools to support these developments.

- Ways in which simulation can operate as a supportive tool with respect to human factors. The field of simulation represents a particular domain of tools available for the support of organisational and technological development.
- The Dutch contribution to this exciting field.

In two respects, therefore, the paper will become more specific:

- It focuses on the interests of people with viewpoints of the second and third kind.
- It focuses on a particular family of tools.

2 Investing in Computer-integrated Manufacturing

After the previous paragraph it will clear that investments in CIM are made for a great number of reasons. These reasons have much to do with the attitude of the company with respect to the production technology it uses. They are also related to the people that prepare or approve of investment decisions. Just as it is practically impossible to buy a car without electronic devices and equipment, it is also almost impossible not to invest in CIM components. The resulting, more or less steady production automation leads in many cases to jumpy and ad hoc attention to the integration of these components, their interfacing, their exchange rules. The fact that most companies are not present today, simply because they never attend conferences like these, has everything to do with this. Of primary interest is the common practice of daily company operations and of the small changes that appear to be involved with every following small step on the automation path. The logic of these small steps never gives rise to the need for reconciliation, for a broader discussion about the direction one takes and for treatment of the fundamental question of whether one does indeed have the proper condition and partners for the longer trip one appears to be making.

For many companies this is not such a stupid way to proceed. There are far too many companies that use this trial-and-error policy productively to cope with technological developments. Investment plans are judged selectively per item taking the budget/costs of each item into consideration and weighing (for example) expected savings, quality improvements to be achieved, labour costs, depreciation period, qualifications, vendor support, software and hardware specifications, etc.

In view of this, companies of Category I have been ruled out for the moment. These companies are not interested in a paper of an organisation researcher and consultant that would be able to help them in structure-seeking or structure-innovating operations.

Let us continue by describing the critical conditions for optimum use of investments in CIM: the success factors for CIM.

On this basis we will be able to precisely pinpoint the need, use and advantages of simulation.

3   Critical Success Factors of CIM: necessary conditions

The implementation of CIM in theory could be regarded as a temporary disturbance of the company's operations. It can then be thought of as a primary process brought to a higher level in a sequence of 'unfreezing', 'changing' and subsequently 'freezing' again.

However, this ideal approach seldom reflects practice. In practice the development and implementation of CIM is a continuously changing process. Let us try to allocate some of the critical success factors to be met in case this continuous process is to be optimally effective.

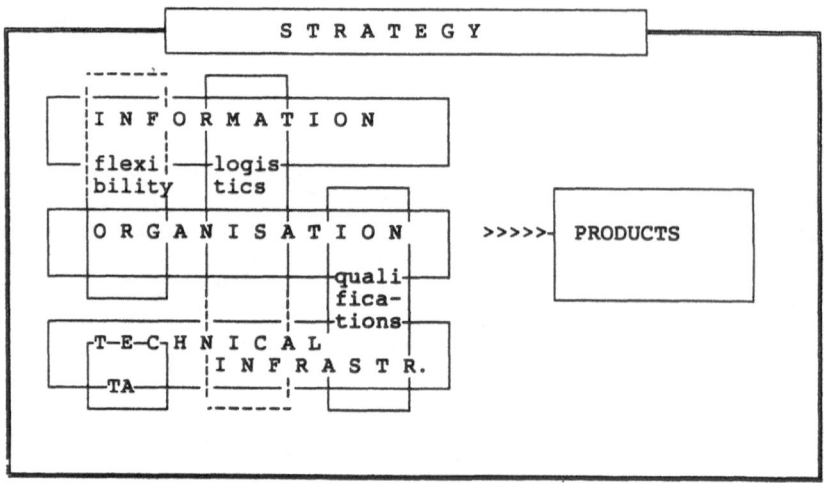

Fig. 1   Model for critical success factors

Critical success factors can, in principle, be allocated to each of the 'boxes' of the model (Fig. 1).

Strategy

Strategic considerations are allocated to the level of the basic position of the company on the 'world' market and the way it can find new ways to address long term challenges.

The more fundamental these considerations or the more complex the problems of the company, the more critical the need to fundamentally change the company's philosophy.

Sometimes the term "World Class Manufacturing" (WCM) is used. WCM, as a philosophy, is composed of the following elements:

- best practice in design engineering, JIT, MRP
- cross functional operations and restructured workplaces
- product and process simplification
- teamwork and the development of human assets
- capability of continuous improvements
- focused factories
- waste elimination
- modern accounting methods and performance measurement

These elements are found to typify the kind of business which is able to compete on world markets and which is characterised by global shifts in competition and market demands.

Another term used to characterise strategic considerations is the quest for a "(Self)Learning Organisation". This concept covers the philosophy that at all levels people are continuously acquiring higher levels of skills and capabilities and also continuously engaged in exchanging the knowledge and information needed to nurture the company's core competencies.

Both terms reflect management philosophies or, so to speak, ideologies. It is not possible to become 'a slightly learning organisation'. Also WCM is composed of the joint approach to all its elements. WCM or Self-Learning Capacities require a fundamental and complicated joint technological, organisational and cultural (second order) change.
The ability to manage this change is a critical success factor in itself. It would require an integral and continuous approach involving the elements we describe below.

Technology

The possibilities with respect to the technical infrastructure must be specified, taking into account the range of available technologies. Only then a company can sensibly select the components that might be useful for its purposes. There are two ways to achieve better assessment of available technologies:

(1)   a company can wait for future technologies available through its vendors
(2)   a company can make an effort at allocating technology assessment and appraisal to specific staff members.

(1) is the more re-active policy, (2) is the more pro-active policy. Successful early innovators have chosen the second policy, knowing that it is a critical success factor to be informed about new developments taking place. Early adapters are aided more and more by training schedules that allow their staff members to gain insight into, for example, European developments presently taking place. Also, their ability to engage in technology cooperation with new suppliers is increasingly supported by all kinds of actions undertaken by the European Commission (VALUE, CIMENET, CIM-EUROPE, etc.).

Logistics

The domain of logistics covers the way the company's internal routing processes respond to efficiency standards and to market demands. Critical success factors include the ability to

- use existing capacity more optimally
- decrease stocks
- increase response time
- achieve quality improvements
- shorten lead times
- shorten delivery times
- increase reliability

The improvements looked for generally are of 'first order' nature. A gradual shift in efficiency is aimed at, without disturbing or fundamentally influencing the primary process.
From 'make-to-stock' one wants to proceed to 'assemble-to-order' and further on to 'make-to-order'. Or, in other terms, gradually from 'maximum use of capacity' on to 'reliability of delivery'.
A critical success factor for improvements in logistics is the development and use of clear indicators. These should allow for measuring a number of variables.

- internal cost/benefit ratios:

    - hours
    - batch sizes
    - raw material
    - personnel
    - machinery
- external customer service level:

    - order lead time
    - realisation of time of delivery
    - distribution
    - flexibility
    - invoicing procedures
    - procedures for complaints
    - order status information: tracking and tracing

Although in most companies improvements in this domain are gradual, in some cases companies have chosen to regard the change process needed as fundamental (regarding its comprehensive approach as a critical success factor). These companies have integrated logistic attention into a change process that also includes technological upgrading, cultural/attitude shifts and structural reorganisation of production. Of course, the specific approach greatly depends on the nature of the problems faced by the company.

Organisation

Attention should be devoted to various parts of the company's structure. The basic question should be whether structures can remain unchanged, whether a slight shift must be aimed at or whether a fundamental turnover is needed. According to a case study by Funk: "(the business) involves numerous factories, design centres, and research laboratories (...) the application of CIM is very complex organisationally. Simply put, the inherent process complexity of manufacture is magnified by this organisational complexity. Clearly, how a firm manages this organisational complexity is a critical factor in the success of CIM."

CSF's for managing the improvement efforts on organisational issues are:

- commitment and pro-active involvement of top-management
- problem solving capacities
- critical review of the company's structure
- internal diffusion of intentions and a visible, participative, management style
- use of informal roles
- breaking through department boundaries and stimulate overlapping of responsibilities
- carefully mixed composition of project team(s)

Information

The exchange and use of information must be at:
- ongoing operations
- logistic sensing
- technology sensing
- change operations taking place
- decisions to be taken

Partly the targets have been dealt with previously under the headings of 'technology', 'logistics' and 'organisation'. Here we need to highlight the following informational success factors:

- the need for public storage areas for relevant and shared information/documents
- the use of oral communication for fast information transfer
- knowledge and information dissemination focused at the core competencies of the company
- the organisation of a distinct communication structure in view of change decisions to be taken

Flexibility

In many cases of implementation of advanced technology, not only in industry and manufacturing, it was found that a flat organisational structure is fundamentally more dynamic than a hierarchical structure. This dynamism is found to be essential for coping with and utilising the principle dynamism offered by modern production technology:

- product mix flexibility
- batch/order size flexibility
- personnel mix flexibility
- personnel volume flexibility
- innovation flexibility

The idea that 'flexible manufacturing' in all these dimensions could be achieved only through installation of flexible technologies has been overruled by the insight that flexible organisational structures are the indispensable counterpart that is also needed, along with highly flexible market-sensing operations.

This means that being able to develop a change strategy towards reduction of structural or hierarchical complexity is a critical success factor. This may sound easy at first hearing, but for most managers this also involves a departure from their basic need for detailed centralised information and control: a true cultural change.

Technology Organisation

A CSF relating to technology and organisation has to do with the system design and implementation process. Cases of CIM implementation have shown that the time needed to develop major system features can best be used at or near the shop floor level. In doing so, the factory is closely involved with the work undertaken by the system engineers or technical consultants. In the various development phases prototypes and system parts can be implemented in close cooperation with the shop floor. This has led in a number of cases to considerably higher acceptance by and also productive contributions of the shop floor.

At the interface between technology and organisation we also find one of the domains of the qualifications of employees.

Employees should not only be able to operate (existing) equipment, they should also be able to basically assess the meaning of equipment under consideration and learn how to use new equipment.

However, taking into account

-       the strategic elements that we described earlier
-       the CSFs on the organisational side
-       the CSFs on logistics and information
-       the CSFs on flexibility

the issue of qualifications of employees will go far beyond the scope of training people how to use existing and new technologies. Especially the German experience has shown considerable changes taking place in the nature and level of qualifications.

Since technological developments are taking place at a rapid pace, the benefits of training in specific technical skills become more and more limited. The 'decay time' of pure technical training is decreasing.

This has been the reason for German industry to invest increasingly in the more general skills that enable the employees to master future technical innovations: a higher level of education and training. Of course the relationship of this to the philosophy of the 'learning organisation' is clear and also the benefits for critical human factors are obvious.

In order to be more precise, we now first need to summarise these critical human factors.

4 Critical Human Factors and prerequisites for tools

In the previous paragraph we cited a great number of critical success factors for CIM. When looking at them closely, we can see the following CSFs that implicitly or explicitly focus on human factors:

- strategic considerations with respect to the whole of the company's operations (production and control systems, organisation and production structure, qualifications, decision-making processes, culture)
- involvement of staff in technology sensing and assessment
- use of clear indicators for measuring key logistic variables and the subsequent use of outcomes
- awareness of customer requirements and customer service level and the cultural change involved
- the need for problem-solving capacities at all levels
- commitment of top management and its participative style
- communication process, information dissemination and decision making
- the need to break down department boundaries and the need for multi-disciplinary teams
- the need to reconsider the companies hierarchical structure
- close relationship of system development to ongoing shop floor operations
- training in terms of machine operations of existing and new equipment
- general 'skilled worker' qualifications

We can draw three subsequent conclusions from this overview:

(1)   Although 'technology' and its proper and targeted assessment and design are vital for CIM, an overview of critical success factors shows that many of them relate in one way or another to human and organisational issues. Some might even say that when it comes to customisation and implementation of CIM, these human and organisational issues are found to be more critical.

(2)   An overview of critical success factors further shows that it would be very difficult to make a clear and lasting division between two or more distinct groups of factors. Critical success factors with respect to each of the dimensions listed intertwine in their actual significance for the implementation of CIM.

(3)   Subsequently, the use of a CIM development strategy in which only experts on specific technical issues take the lead will be less effective than a strategy that stresses consultation and participation on the part of those involved.

This inevitably means that any effort to smoothly bridge the gap between the present state and the future state of operations of a company always needs to look at technical as well as human and organisational issues.
Supportive tools in this respect are:

- advanced analytical tools to picture the present state of the company and its market
- structured methodologies and clear sequences of concurrent steps to follow in the design, development and system implementation process
- means/tools to overcome the 'mental block' that exists due to the fact that one needs to imagine features and effects of not yet installed equipment and structures.

In the next paragraph we will be dealing with the particular family of simulation tools. Simulation is especially interesting because it holds the promise of being able to support the design, restructuring and training process on all three domains.

5   Definition of simulation

For the sake of clarity we first have to prevent possible confusion between business simulations and business games. Both terms are used to represent simulation models in which participants fulfill functions and interact as participants in a modelled setting.
The difference between the two has to do with the fact that in business games a computer model is used to represent the actions and decisions of employees that are not present in the model in real life. Effects of decisions taken at the board level are calculated by computer and result from exercises that the 'hidden' computer model carries out on the data and on the basis of assumptions about employee behaviour, etc. This means that business games are always subject to a struggle between the assumptions of the model's designer and the reality of individual companies. They never match and therefore the outcome of business games will always be an artefact whose validity still needs to be tested.
In business simulations the actions of employees in relevant departments or at relevant levels are a natural part of the model since these actions take also place within the reality of the simulation setting. Real-life participants undertake actions and complete production, planning, information, control and other tasks.

In order to be able to judge impact and usefulness, we should note here that any simulation must contain valid and integrated representations of technological, human and organisational issues. The reason for this is that, as we have seen in the previous sections, in real life these dimensions cannot and should not be separated. The 'problem' of games is that they, by definition, contain a more limited representation of real-life conditions.

In this paper we shall not deal with that other member of the 'simulation family': computer simulations. These simulations are meant for specific optimisation purposes of system components or of the total system. They can never incorporate the human and social organisation dimension. However, they do have a particular function that cannot be ruled out as being irrelevant.

6   Advantages of simulation

The advantage of simulations is in the availability of inexpensive, comprehensive and in-depth analysis, training and change models. This almost sounds like a dealer's sales talk, so we need to specify.

Analysis

Simulations allow us to analyse basic social organisation features of the company. This can be done in two ways:

1.   by designing a valid model of the company (and its environment) and to investigate, through repeated simulations, the structural characteristics of the company
2.   by inviting people of the company to participate together in a simulated organisation.

The first option is more research-oriented. It is especially suited for highly sensitive problems that need very deep analysis before any measure can be taken. The use of various research tools allows one to study processes and structures that develop under simulated conditions. The very fact that repeated modelling and replication are possible creates the opportunity to really pinpoint basic and essential characteristics.

The second option is action-learning-oriented. It allows members of the company to examine their mutual social-dynamics and organising processes as they in fact appear to be. Participation in a simulation model first leads to insight into the processes as they are in the simulated setting. Subsequently, together with the participants an assessment is made of whether these simulated relations in essence mirror the actual (real-life) process, including its positive and negative features. The fact that in a simulation the positions of the respective participants can be turned upside-down means the occurrence of "taking-the-role-of-the-other" effects. People come to understand deeply why the organisation functions as it does because they have been able to experience what it is like to be in another position. This insight is needed for organisational development.

Methodological steps

Taking lectures on information technology is one thing, really knowing what is meant with the concepts in a specific setting, involving diverse disciplines, is something quite different. As a first methodological step, execution of simulations contributes considerably to the development and common use of a shared vocabulary to understand and jointly further develop the interfacing between technology and organisation.

Secondly, simulations facilitate to-the-point involvement of key-users in the process of system specification. The fact that a valid simulation incorporates technical as well as organisational characteristics allows key users or different levels to experience and work out specifications in a mutual and comprehensive effort.

Thirdly, simulations make it possible to use and experiment with not yet installed systems, procedures, structures and control mechanisms. This increases the awareness of the technical possibilities (experimentation and prototyping) and their likely implications at several levels of the organisation.
This is linked to the possibility of increasing awareness and assessment of alternatives for organisational structures and procedures.

Fourthly, simulations represent an in-depth training setup for the preparation and training of employees regarding how to approach new technology, new structure and new job content in a joint effort. Especially repeated simulations at various stages of the design and development process increase involvement of personnel considerably.

Overcoming 'mental blocks'

Most stages in production innovation processes demand imagination from the people involved. This imagination is needed for full support in the CIM development process, including its repercussions for organisational and cultural changes. Since people are used (and tied) to existing technologies, procedures and routines, it is hard to imagine new technologies and their possible use and effects. The problem with many (government-sponsored) demonstration programmes is that they give insight into someone else's solutions. This leads to mental blocks preventing people from fully understanding the possibilities and implications for their own situation.
However, in CIM implementation, assessment of user requirements is needed. Sequenced and repeated onsite testing is needed. Prototypes need to be evaluated, not only in a laboratory environment, but in a real life (or close to real life) situation. Simulation could be of use here.

7   The Dutch Contribution

The challenge has been to have a simulation setting in which the full scope of CIM issues can be represented

- on a valid basis
- so that they are suitable for use in diverse companies

Such a simulation setting would have to:

-       incorporate valid (real-life) social organisation as well as technical characteristics
-       flexibly attune the basic organisational model to the variety of companies it would have to support
-       flexible attune the production technology used in the simulation to the existing and future production technology of these companies
-       flexibly match the specific developmental steps that are taken in individual companies

Our Exploratory Action within the scope of the Esprit programme (5603) has, among other things, dealt with the question of whether such a simulation facility exists anywhere in the world. It does not.

What does exist is a number of highly relevant elements and examples of simulation practice that together would lead to such an ideal simulation setting.

Promising developments have taken place in the Netherlands with respect to one of these elements. This is the area of simulations of complex organisational processes. The people working on this area come from Erasmus University, the Technical University of Twente and from InterVisie.

In the first place, we have been able to develop an approach to design valid simulation models of organisational processes and structures.

In the second place, we have been able to use these models for in-depth analysis of complex organisational and social problems. In the third place, these models have proven to be extremely effective training and organisational change tools.

In the fourth place, we have been able to support the development of organisation theory through research and experimentation on the basis of simulation.

Basic features of these models are that the participants are handed instructions about the organisational structure with which they begin and about their position in it. The organisation is provided with a very brief and neutral 'history' of earlier production and sales.

The simulations are composed of 5 to 10 succeeding simulation years of 60 to 90 minutes each. These years are not interrupted because a computer model would have to calculate the effects of decisions taken (as is the case with 'games' and some other simulations). The years succeed each other without interruption, as is the case in real life.

Another general characteristic is that all essential functions are represented in the models. There are no functions left out or represented through the calculations of a computer model.

Participants start their work in the simulated setting much the same as if they had just started working in a new job.

Only the time scale is reduced and some non-essential features have been left out (in order to achieve reductions in time).

The result of this is that participants are in complete control of what they do. The researchers or consultants do not intervene in any way after the start of the simulation. The organisation really comes 'from' the participants. Everything they do is on their own initiative and they are themselves responsible for the results (in terms of income/turnover/profit as well as organisational structure, effects of decisions, procedures, plans, social and labour relations, etc).

This means that these simulations are not subject to the disadvantage of other simulations, i.e. their 'experiential demand'. This says that the process within and the outcome of most simulations (and games) depends on influences of the model's designer or of the game's coordinator.

The problem of this experiential demand is that one can never be certain whether the outcome is realistic. The outcome might as well be an artefact of the model itself.

Why is this so important?

The outcome must be realistic (valid) in order to be sure about the measures to be taken in real life in order to be certain about the effects of simulated structures and systems.
Especially with social processes and organisational issues, it would be disastrous if one were to take measures or set up structures which after some time prove to be completely wrong because they were based on untrue assumptions.

The simulation approach in the Netherlands is based on the challenge to develop models that avoid this problem. The problem can be avoided by specific validity restraints.

The validity of models is composed of:

-       Their level of 'psychological reality'. Within a few minutes after the start of a simulation participants should act and think in terms of the simulation. This means that they occupy a position in the simulated organisation and act as if this position were their real occupation.
-       Their structural validity. The organisational structure that participants develop out of the 'offers' of the starting situation must resemble those of the real-life situation. Preferably after one simulation year and checked by repeated testing halfway and at the end of the simulation.
-       Their process-related validity. The organisational processes and the issues taken up by participants must resemble the processes and issues at stake in real life.
-       Their systematic validity. The structures and processes detected in every experiment with a particular model must be the same, despite changing groups of participants and despite various locations in which the simulation is executed.

In addition to these general demands with respect to validity, a simulation model should be able to predict phenomena that have not yet occurred in real life. Especially when using a simulation to experiment with system features or management measures, one should be able to say with a fair amount of certainty that effects within the simulation will or would also occur in real-life.
Of course, this is a characteristic that needs proof through 'repetition'. However, at this moment, let us leave the issue of forecasting and prediction and its methodological implications for what it is.

To summarise, in the Netherlands there is a growing experience with simulations that in principle meet the demands of advanced technology management that incorporates focus on organisational and human issues.

Further action needs to be taken on the following:

(1)
The first element to further expand upon is that of making it possible to use advanced production equipment in these kinds of simulation models. One can reduce the time scale, reduce essential organisational characteristics and reduce

complex processes in a valid way. Is it also possible to use manufacturing or CIM components in such simulation settings?

(2)
The second element to expand upon is that of creating in very short time the basic situation for a particular problem and company. This would mean the production of functional descriptions and of the general organisational structure with which the participants start the simulation.

The prospects are hopeful:

In the first place because computer simulations of equipment allow for a realistic, simulated, production and planning process. The use of PLCs, coupled to this simulated equipment, enables simulation of a CIM environment (including CAD, PPS, CAE, CAPP, etc.). This means that an effective simulation could very well consist of integrated computers as well as social modelling.

In the second place the developments with respect to artificial intelligence permit the use of expert systems that produce the starting initial conditions for an endless row of specified settings/models aimed at any likely company.

The real challenge hereafter would be to integrate simulations of this kind into a 'CIM Development Methodology' or CDM to be used to comprehensively structure the design, development and implementation process, while taking local industrial needs, demands and specifications into account.

Joint Project: Simulating Work and Technology - Development, Implementation and Application of Computer-Based Systems for Planning, Simulation and Animation in Manufacturing

Prof. Dr. Friedrich Wilhelm Bruns; Lüder Busekros; Axel Heimbucher
artec, University Bremen
Bremen Germany

1   Introduction

The project "Simulating Work and Technology..." is a cooperative research project of the School of Technology of Bremen and the University of Bremen sponsored by the Bremen State Program Work and Technology. The project aims at analyzing the potential of simulation as a method of experimental system design thereby particularly emphasising

- alternative perspectives,
- technological assessment,
- participation.

As a result of the research project recommendations for the design and the use of simulation systems, they will be oriented to

- the recommendation of the VDI "Social Oriented Design of Automation Projects" (VDI 89) and
- the Human-Centered Methods (HCM) of the Danish research group of work and technology of Rasmussen and Laessoe (Laessoe 89).

Both these approaches are characterised by a more or less strong accentuation of human, social and ecological as opposed to purely technical aspects to be found in many automation projects.

2   Project Description

The simulation technique supports the modelling of certain aspects of reality, experimentation with these models, and evaluation of the experiments with the aim of gaining a better understanding of reality, of learning to handle it in a better way or of changing it. Simulation is to serve as a means of analyzing and assessing alternative forms of technology and organisation. We start with the hypothesis that the user's perspective and interests have great impact on the phases of model building, experimentation and evaluation. That means: simulation is a highly subjective task. This subjectivity should be revealed and put up for discussion among the persons involved in the simulation process.

Referring to their functionality (what can be modelled?) and their usability (which level of abstraction and which focus is selected?), simulators which are available and in use up to now primarily reflect the perspective of planners and system designers more than the participation of those persons who have to work in the future system.

The aim of the research project is to analyze existing simulators with respect to their suitability for incorporating a greater variety of perspectives and a more general usability. Suitable simulators have been selected and are to be improved in an iterative process including the following steps

1. Training of system designers (planners, engineers) and persons involved (workers councils, users).

2. Application of simulation in order to integrate new organisational structures and processes into existing production structures.

3. Improvement of simulation tools.

The the project is oriented to central demands of current research in work and technology as formulated by the design recommendation of the VDI and of the Danish HCM group. These central demands are

-        giving technology a greater capability of negotiating functions and interests among those who are affected by technological developments,
-        identification of alternative possibilities of technological, organisational and social realities,
-        support of socio-technological experiments for technological assessment,

The design recommendation of the VDI and the Danish HCM approach are briefly outlined as follows.

The VDI advises planners, users, managers and workers councils to shape man-machine systems in a humane and economical manner based on the following principles

-        simultaneous planning of humane work and technology,
-        cooperation between planners, designer and users at an early stage and in a continuous way,
-        consideration of the strategy of the enterprise
-        participation of employees,
-        integration of persons involved
-        extensive information and training,
-        consideration of market, society and environment.

A particular feature of the methodological procedure is to be seen in the iteratively performed design process comprising a planning phase, a conceptual phase, an implementation phase and a test phase. Performed with the participation of the persons involved, this design process provides simultaneous elaboration of two design philosophies:

a work-oriented approach conceiving work as qualified and motivated as possible under optimal organisation and shaping of work and
a technology-oriented approach conceiving different possibilities of automation.

Design circles, possibly comprising managers, users, developers, planners and workers councils, cooperate during the design process. In order to better realise and estimate the effects of automation on man and environment, the design circles discuss key questions related to work conditions/work organisation, personality development, skills, user protection, general effects (employment, national economy, nature, society) and the basic conditions of business management.

With their methodological approach of cooperation between researchers/developers and planners/users Laessoe and Rasmussen follow the research tradition of the Scandinavian "Collective Resource" and "Action Research". They clearly distinguish their HC approach from socio-technical approaches which aim at equilibrium between technical and human perspectives. Instead man and his interactive possibilities is seen as the basis for the development of production techniques and the organisation of work.

Some of the methodological principles of their approach are as follows:

- experimental prototyping,
- active participation of users in discussing and formulating the goals and means in the planning phase and in the implementation,
- rejection of the ambition to establish an external objective relationship to the object of investigation; instead establishing a subjective relation between researcher and research object,
- cooperative problem definition while revealing the different interests and motivations, world views, perspectives and concepts,
- realising the importance of dialogue between different concepts,
- design-by-doing,
- reciprocity versus model power: "Do the researchers' better working conditions and greater experience with project work imply that the cognitive process is constructed around their system of interpretation (models) and methods of work? Or is it possible to successfully create organisational frames and find methods for cooperation that ensures a reciprocal process?" (Laessoe 89, S. 48)

Special potential of simulation

Simulation represents a special potential for the design of work and technology, oriented to the needs of users involved, which has not been realised sufficiently up to now.
First of all simulation is a representation of certain aspects of reality in a model. This simulation is a means of formalisation similar to the development of software. It contains a description language for complex parallel distributed processes. Moreover, simulation is used with the aim of gaining knowledge about the dynamics of a system and of optimising it.

Secondly, simulation is a method for experimentally dealing with uncertainties. Here we see its significance. If simulation is understood as the iteratively performed cycle of model building, experimentation, interpretation and change, in which all the phases are highly subjective and should therefore contribute to critical discussion of different interests, then there is still great potential to be tapped.

The method itself contains some interesting concepts of software engineering:

- a software structure which is not oriented to a central sequential flow but to distributed objects with individual and hereditary attributes and their message oriented relations (object orientation)
- a shell that supports an experimental, cyclical process suitable for quick changes (rapid prototyping),
- a data structure which allows for dealing with uncertainties and state and event probabilities.

There is still much to do to broaden these existing concepts and to make them available. Our project attempts to contribute to this aim by elaborating the following items:

- differentiation and explication of the objectives of simulation regarding the following: orientation to technical, organisational training, to systemic work as well as to economic and political aspects,
- a more conscious consideration of the model limits, the relation of man, system and model,
- a stronger emphasis on possible changes in perspective, view of the entire system from above and of single aspects from below, concretely, visually, abstractly and statistically, in real time and distorted, roughly and in detail,
- usability as dialogue support in the discourse of differently involved and interested persons,
- an experience-based modelling tool which requires less abstract thinking by means of symbols (a handy construction kit for model building).

3 Project Plan

Foundation of a pressure group: simulation applications,
Rough information about 50 simulation systems,
Rough catalogue of criteria for selection,
Selection workshop: system presentation by 8 invited distributors,
Purchase and installation of three systems (SLAM, AUTOMOD, SIMFLEX),
Training on the basis of published case studies,
Development of a course concept,
Courses for the use of simulation as an analytical tool for engineers, foremen, skilled workers and students,
Industrial application of simulation: gear quality control, food production, harbour planning,
Use of simulation as a means for communication and discussion,
Courses for workers councils, planning staff and users,
Criteria for an improvement of simulation systems regarding user participation in

the design of working conditions,
Prototyping of simulation modules.

4 Initial Experience

During the first year of project work we gained the following experiences:

1. The potential of simulation has not yet been realised sufficiently by possible users.

2. The basics of simulation which we introduced in our courses, although rather abstract, were understood by the participants with different previous knowledge and led to autonomous model building processes within the groups.

3. Already the act of modelling was highly discursive and uncovered different interests and perspectives.

4. Observed deficiencies of simulation systems:

- rigid flow oriented links between objects,
- only local flexibility in a global determinism, no local reaction to remote events or states,
- little possibility of interaction, no possibility of model changing during experimentation,
- no possibility of time inversion,
- lack of concreteness concerning different model views,
- few possibilities of changing perspectives,
- no satisfactory support for entire simulation cycle (modelling, experimentation, interpretation, change),

5. The use of a physical construction kit for experience-based modelling is especially suited as a first step of the dynamic simulation model. Furthermore, a way has to be found to transfer the physical premodel to the dynamic computer model.

5 Further Questions

1. How are the quantitative aspects of the simulation related to qualitative aspects of products, technology and work?

2. Is simulation mainly an instrument of optimisation and control aimed at the reduction of scope and intensification of work? What kind of alternatives are possible?

3. Can simulation be used to turn the dominant system-oriented view upside down into an individual perspective?

4. Can it be expected that with the help of simulation the employees will perform the job of management? Should they in the first place?

5. How can simulation be integrated into real systems as a tool for experimentally looking ahead in order to get more insight into real system behaviour?

Appendix

Selection Criteria for a simulator of production systems

functionality
        alternative modelling
                continuous
                discrete
                        time driven
                        event driven
                                activity-oriented
                                process-oriented
                                transaction-oriented
        stochastic distributions
        conflict handling
        configurable control strategies
        model-optimising tools
        strategies of evolution
        medial continuity for the entire simulation cycle
        integrated tools for analysis
modularity
        integration of subnets
        module library
hierarchy
openness, flexibility
        module development
        extendable functionality
        user interface
        programming interface
                C, FORTRAN
        interfaces to data base management systems
        interfaces to expert systems
user orientation
    adaptability
        choice of abstraction and view
                networks
                decision tables
                sequence tables
                languages
        breakpoints
        backtracking
        support of series of experiments
        statistics
        financial aspects
        visualisation
interfaces
        CAD (layout)
        CAE (topology and kinetics)
        CAP (workplan and workloads)
        failure input
        animation
        realtime processes
        user interaction
portability
        Windows, OS/2, Unix
possibility for further development
        source code access
        documentation
        programming interface

support

Literature

Laessoe, J., Rasmussen, L. B.: Human-Centered Methods - development of Computer-Aided work processes. Esprit-project 1217(1199). Human-Centered CIM systems, Deliverable R18. Institut for Samfundsfag, Danmarks Tekniske Hojskole, 1989

Verein Deutscher Ingenieure VDI: Handlungsempfehlung: Sozialverträgliche Gestaltung von Automatisierungsvorhaben. VDI, Düsseldorf, 1989

Bruns, F. W., Busekros, L., Heimbucher, A., Rocker, W., Scheel, J.:
Auswahlbericht fber anwendergerechte Simulationssoftware inklusive Entscheidungskriterien. Zwischenbericht an den Projektträger Arbeit und Technik des Landes Bremen, 1991

FPA training within the Dutch apprenticeship system

D. van der Ent; Peter de Jong
SOM Stichting Opleidingen Metaal
Woerden Holland

1 FPA TRAINING WITHIN THE APPRENTICESHIP SYSTEM

1.1 INTRODUCTION

The SOM Group is the national institute for training courses in the metal industry in Holland. The courses offer a range from vocational training within the scope of the apprenticeship system to refresher and entrepreneur training courses. The apprenticeship system provides vocational training for people aged 16 and over (more than 12,000 each year). Practice is the central element of these courses. A dual training method is used. On the one hand, trainees receive theoretical training at a day school. In addition they acquire practical skills within companies or at institutes for practical training under the supervision of experienced craftsmen.

1.2 FPA IN THE DUTCH SITUATION

This paper deals with FPA. The meaning given to it by SOM will be explained below.
SOM's working definition of FPA is: "the integral management and control of a production process whose input (raw materials and semi-finished products) and output (semi-finished and finished products) are variable."
Consequently, the emphasis lies on process control. This control affects all functions in a company. FPA is multi-disciplinary and makes demands on all of the staff.

As regards the implementation of FPA systems, Holland is by no means behind when compared to other countries. In Dutch industry FPA applications are most widely used in the metal sector.
The adoption of FPA is usually a gradual process. Automation processes generally create 'island' solutions in the sphere of CAD and CNC applications. These (technical) 'islands' create the need for integral control of the flow of goods and stocks. Industries are increasingly becoming aware of its importance, with special attention being given to logistics. Further integration will require, amongst other things, good communication lines between the islands. As a result, communication will have to be standardised.
These developments make higher and higher demands on the information system and the people operating it. Nearly all industries have partly automated systems for

production control. But these systems rarely allow further integration. Although industries are still divided among themselves as to rate and manner of introduction, tentative steps are being taken towards an overall FPA solution. The majority of software developments are concerned with the CAD/CAM combination and with flexible production control systems.

1.3   FPA VERSUS MAN

FPA applications are often directly connected with the fully automated factory equipped with robots and featuring fully automated product assembly and control. This concept was put forward as the factory of the future, as late as in the 1970s. During the 1980s efforts were made to put this concept into practice. It turned out to be a total failure for General Motors in America, and results at Volkswagen were not very encouraging to continue the "Halle 54" project either. Japan was more successful with a small number of projects.

All 'experiments' indicate that even in companies where automation processes have reached high application levels man remains the chief link.
The current situation with respect to FPA applications shows that the option is mainly suitable for hybrid solutions: systems including both automated production processes and partial solutions with a balanced interplay of man, computer and machine.
The option is a hybrid solution allowing both computer and man to control the system.

The end of the 1980s saw the worldwide start of the creation of a Human Centered CIM concept. In Europe relevant research is being done within the EEC COMETT project. This concept puts man at the centre. The concept aims at creating a production environment having relevant instruments and customised to human standards. It employs new production concepts and software approaches. One of the instruments is training, which is the very reason for SOM to participate.

Recently, the view has been expressed more and more clearly from different sources that the social and organisational aspects of the project are as important regarding successful implementation as is the technical aspect.
It appears, for instance, that flexibility is not so much determined by technology as by man, assisted by computerisation and automation.
The flexibility of the organisation and the quality of production are determined by communication, whether or not formalised, between people themselves as well as between people and machines.

Nationally and internationally there is a growing awareness that personnel training is an essential factor in the implementation and employment of FPA systems.

Initiation and implementation of the FPA approach in industry and, particularly in the MKB are still matters mostly dealt with by the higher echelons in industry.
The creation of training facilities on behalf of the lower echelons might make the awareness of FPA in industry more acute.

2   THE SOM APPROACH TO FPA TRAINING

2.1   TRAINING PLAN

2.1.1   TARGET GROUP

The target group includes officers dealing with planning, preparing and controlling production and/or assembly. They are the links between management and production. They possess sufficient innovation potential with regard to their jobs and are used to employing automation as an instrument. These jobs are characterised by a well-balanced blending of socio-organisational and technological aspects that are needed to implement FPA and to work within its scope.

Middle management functions make up the principal part of the training target group. Some examples: process planners, engineering draughtsmen, cost accountants, planners, warehouse managers, departmental management of the factory departments, (CNC) operations foremen, quality assurance personnel, etc.
This target group may be active in a variety of areas, e.g. machining, assembly/maintenance, engineering, aircraft construction, agricultural implement construction, etc.

2.1.2   TRAINING ROUTE

The participants consist of people with a certificate of the secondary education level, such as senior technical and vocational training schools, coming from various disciplines and having various levels of experience.

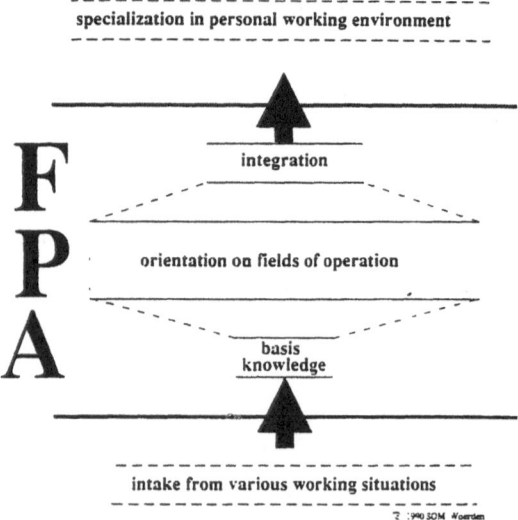

Fig. 1   The FPA training route

First a certain amount of basic knowledge will have to be assimilated to clarify the subject matter presented during the course and to effect a uniform level of knowledge as the group will often have to work together.

This basic knowledge will be enlarged during the treatment of the individual fields of operation.

During the integration part of the course the interaction between the fields of operation will be explained. It is estimated that this part will up one-third of the training.

If, after completion of the FPA training, further schooling should be required, specialisation will be called for in the FPA application in the personal working environment/situation. The extent to which SOM will be able to meet this demand is still under discussion.

2.1.3 ORDER OF TREATMENT

In dealing with the various modules, the best order to be followed would be the one which, product-wise, is best suited to the practical situation: from design to production and inspection.

1.Basis	Basic knowledge
2.CAD/CAM 1 and 2 3.CAC 4.CAPP 5.MP&C 6.CAQ	design calculation (estimating and costing) process planning logistics (planning) quality assurance
7.FPA-I	integration (cooperation)

Fig. 2 The order of treatment

3.2.4 DURATION AND CYCLE TIME

The 1-year vocational training consists of 40 weeks with a 10-hour curriculum weekly. Three hours are devoted to Reading Drawings, Science, Communicative Skills and Personal and Social Education.
The technical part will effectively take up approx. 210 hours.

2.2 TRAINING MODEL : THE TRAINING WORKSHOP/FACTORY CONCEPT

2.2.1 GENERAL

As referred to above, the FPA training must be formative in terms of

Modules		Cycle time in weeks
basis		3
CAD/CAM		4
cad/cam1	(2)	
cad/cam2	(2)	
CAC		3
CAPP		3
MP&C		6
CAQ		3
FPA-I		12
		total 34 weeks

Fig. 3   The duration and cycle time

- flexibility
- integration
- interaction, etc.

This objective can be met by combining technical training with practical exercise. Through actual experience, the student acquires an understanding of the cohesion of information systems and causes and results of operations done. To impart a sense of realism, all company functions and their interactions should be simulated. To this end a scale model is used. In this way we avoid the impression to be working on an ideal laboratory model. Attention must stay focused on the basic subject matter. SOM intends to realise this in a Training Workshop or Training Factory.

The Training Workshop/Factory could be described as a company where students occupy various jobs in a job rotation system. The various jobs are realised in a manual and automated hybrid model. In the training workshop/factory, the emphasis is on control and related computerisation of a production/assembly system. In spite of its simplification, this model must represent the aspects of the control complexity occurring in MKB, but also in business units of large companies.
Material transformation and shop floor automation play secondary roles.

A number of sub-models can be distinguished in defining the make-up of the training workshop/factory.
The following models must be defined:

Physical model: What products are manufactured by what production means?

Organisation model:	Who organises this model and how?
Information system:	What information is to be exchanged and with whom?
Automation model:	What operations are done by computer and what type of data processing is automated?
Training strategy:	How is this to be explained in a course?

2.2.2   TRAINING WORKSHOP/FACTORY GROWTH MODEL

It is important for the scale model to grow along with the developments in industry. This means that the scale model must have a modular structure so as to easily accept innovation in specific areas. The training workshop/factory concept is a growth model, allowing for increasing automation, both in production planning and production/assembly. Of course, it is also possible to make the change from an automated to a more manual production process if developments in industry pointed in that direction.

2.2.3   AUTOMATION MODEL

How far should one go in automating an educational situation and what should the objective be?
It has been SOM's deliberate choice not to employ a fully-automated production line with PLC, robots, CNC machines, etc. to prevent a focusing of attention, consciously or unconsciously, on technology and related problems. It is necessary for trainees to learn how to function in a system, not to learn how to adjust a system as a whole.

Partially-automated production is desirable as:

- it is most like the industrial situation in Holland;
- the activities (whether or not automated) have to be tested in order to be able to recognise problems in automation and integration processes.

Automation is an essential ingredient of FPA and, as a result, of training. In the scale model, therefore, an island automation process is carried out for each company function or field of operation. Interaction between company functions is not to be fully automated. For an explanation of principles, man must act as the go-between.
The student must have a clear overview of how the principles work when automation is extended within the framework of the growth model. It must be tested again and again to see whether attention for technological aspects still complies with the orientational and formative objectives.

2.2.4   INFORMATION SYSTEM AND ORGANISATION MODEL

In the information system the information flows and the functioning structure of the training workshop/factory are laid down. In terms of production planning and production/assembly this means that all input and output of data as well as flows of material are defined.

The organisation model describes the various functions with related tasks in the training workshop/factory. Furthermore, procedures related to the completion of tasks are indicated together with relevant responsibilities. The number of these tasks will, of course, have to remain limited and be pared down to essentials, without losing their representational character.
SOM has made preliminary analyses of organisation models representative of the metal industry. Basic models have been worked out in detail for two company typologies:

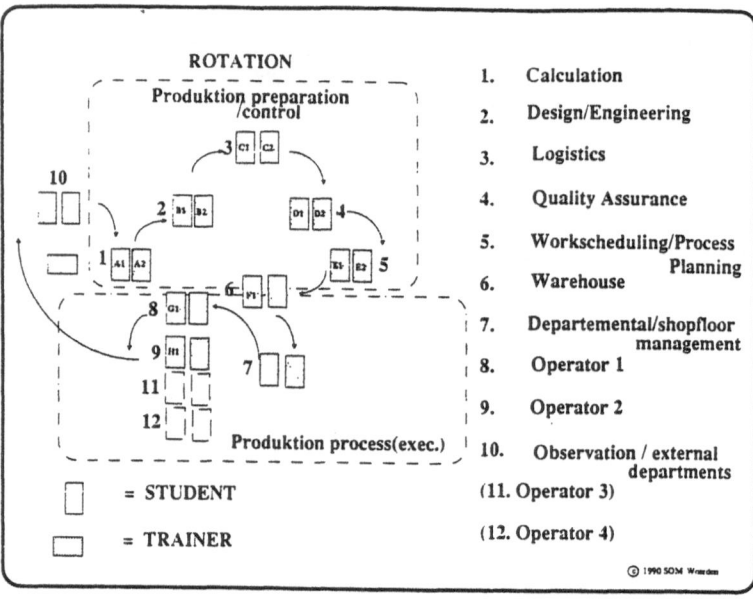

Fig. 4   Roles/functions within the model

order-controlled discrete production
stock-controlled discrete production

For further elaboration of the basic models in terms of an information(flow) model (more detailed) it has been decided to develop a hybrid of the above typologies:

program-controlled discrete production.

Ten or twelve workstations are to be distinguished, accommodating eight functions.

Functions 1 to 6 are functions with direct reference to a field of operation.

Functions 7 to 9 are organisational roles on the shop floor.

Function 10   is a role with a dual function:

1.   Observation of the (integral) whole; didactic function.
2.   Representation of functions/roles not included in the model: input/output of, for example, purchasing, sales, production management

2.2.5   PHYSICAL MODEL

The physical model (system) can be implemented in various ways. The minimum configuration consists of an island for automated production planning and control, and a workshop for production/assembly with conventional processing machines and a (slightly-automated) assembly line. This configuration can develop into a fully-automated environment.

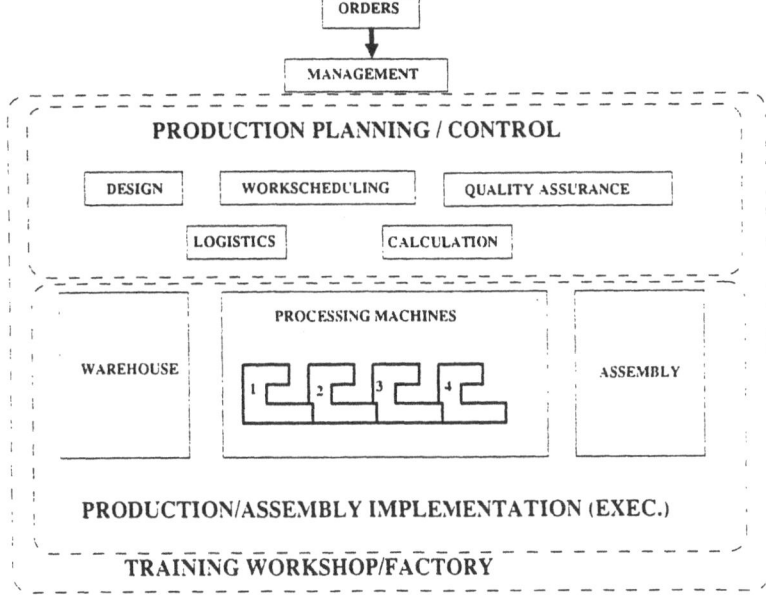

Fig. 5   The production planing/control system

The configuration as a whole will depend on the selected product, product structure and product routing. Manufacture of a variety of products, such as measuring rods and castor wheels, is currently under consideration.

The configuration can be like the one shown below. It would require the training workshop/factory to be situated in a large room.

The computer configuration is to consist of a network of nine computers:

- 5 AT computers for production planning
- 3 AT computers for production
- 1 386 computer as network server to be operated by the instructor.

2.2.6 TRAINING STRATEGY

Following is a summary of views and conclusions serving as starting points for the training strategy:

- the training must be formative through gathering experience.

    reality must be simplified whilst remaining representative.

- the form of training (for FPA integration) is to be a simulation game, with each student carrying out each function at least once in a job rotation system.

- The use of CAT/IAV (Computer Assisted Training/Interactive Video) is important with respect to:
    - showing the relation with the actual situation.
    - learning the (function) role in the game.
    - educational efficiency and didactic efforts.
    - economic reasons (lower investment in expensive machines and other equipment).

■ the scale model must explain a complete, operating factory, without becoming the focus of attention itself. The training workshop/factory is not to be a perfect laboratory model. Evaluation of the model must be avoided. It is important that there is a transfer to personal situations throughout the training process. This can be stimulated by a good model.

PLAYING THE SIMULATION GAME

Students are allotted their places according to the Figure below:

max. number of participants 16
min. number of participants 8

First they have to study their roles. They have already acquainted themselves with the contents and peripheral conditions of their functions in the previous models. This information on the role in the game will have to be presented via CAT (Computer Assisted Training). While the game is being played, it must also be possible to consult this CAT.

Now the game can be started. A initial situation must be simulated to enable everybody to contribute.

The game is to include a number of breaks to point out principles, for reasons of feedback, or to be able to start anew.

JOB ROTATION

The system of job rotation has been chosen to have each student occupy each place once (cycle time 10 weeks).
A second round (cycle time 3 weeks) has been included to enable the students to assimilate the experience they gathered in the first round.
Points of consideration:

- each student is to go through all the stages once more so that he or she will be able to use the experience gathered in a second round;

- to have each student go through a certain number of prescribed cases on a fixed location of personal choice so as to optimise the process as a team. Comparison with other groups might encourage competition.

TEACHER'S ROLE

In contrast to current teaching practice, the teacher will be required to provide an intensively participatory approach and guidance.
He/she will have to manage and control the simulation game and supply the necessary input. The simulation game in the training factory will, of course, be on a time-schedule. Dependent on progress and group composition, the teacher will have to adopt a flexible approach towards the programme. If participation is insufficient, the teacher may act out the lacking roles with the necessary help. He or she is the 'manager' in each game. The teacher's tasks include network control.

CONTROL AND TESTS

Integration model

The teacher will have to be provided with good instruments and indicators to be able to measure and register progress at an individual level as well as at a team level. This is an essential aspect to the development of the simulation game and the related study management system.
A number of standard cases with clearly-defined, fairly standard problems and solutions must be made recognizable during the game.

Other modules

Theory and practice will be tested separately.
The practical aspects will be tested by means of reporting on the exercises/projects carried out.

Practice-oriented tasks

Tasks are also to be carried out in the company that employs the student. These tasks will consist of analysis/research related to the subjects dealt with during training.

These practice-oriented tasks must become an essential part of the training in order to enhance the abstract faculty and to promote knowledge transfer. Moreover, (the monitor of) the company will be more able and encouraged to give guidance when the interaction between training and company is recognizable.

Integrating Organisational Development and Learning Development in a CIM Context

Barry Nyhan
EUROTECNET
Bruxelles Belgium

1   Introduction

Advanced technological organisations and the people working in them need a new attitude to learning. This situation has come about because the "knowledge" required by companies in today's turbulent business environment dominated by changing markets and improvements in technological control and production methods is becoming more complex.

"Rule" based knowledge is no longer adequate. The knowledge required today must be based on "understanding" complex situations and the ability to implement appropriate decisions. This requires the ability to read situations and make a human judgement based on existing knowledge, taking all of the present socio-technical factors into account. This requires a new kind of continuous and dynamic learning, based on past understanding and incorporating presents insights. People today more than ever need the ability to actively learn for themselves

EUROTECNET is a European Community Programme aimed at promoting innovation in training systems to respond to technological change. One of the areas which EUROTECNET focused on was the fostering of people's ability to learn. The Self Learning Competency (SLC) idea was developed therefore as an action-research project. SLC enables people to actively learn in a variety of circumstances throughout their lives. It can be defined as "an active power within people, making them engage continuously with all of their experiences (in an open and inquiring way), to understand and master them" (EUROTECNET paper, Nyhan, 1989). The word "self" in Self-Learning Competency refers to the need for independently driven learning so it must be interpreted in a social as well as an individual sense. Heidack (1989) writes about the concept of "Cooperative Self-Learning"

As well as drawing on the conceptual work and the conclusions which emerged from the SLC Enquiry (Nyhan,1991), this paper also utilises other EUROTECNET findings dealing with the effective implementation of CIM. The paper first examines the broad organisational context within which a new kind of learning paradigm is required today. Then it looks at the active learning environment necessary to fully exploit modern Computer Manufacturing Systems.

2 New Organisational Paradigm : New Learning Paradigm

In 1978 Argryis and Schon attempted to answer the question in their book on organisational learning, "How does an organisation have to be structured so it can learn?"

They drew a distinction between "single loop learning", which means detecting errors and correcting them through changing strategies and tactics, and "double loop learning", which means not merely detecting and correcting errors, but restructuring the norms and guiding principles which direct the organisation. Political changes in Europe in the West and the East, the globalisation of world markets and the globalisation of high technology production itself (Henderson, 1989) increased competition between major trading blocks, growing demands by consumers for individually designed products call for new norms in organisational design and development. Advances made in new technology are now being applied to all of these facets of business, thus creating high-tech manufacturing organisations.

The hallmark of the new organisation is "Flexible Integration". The company must be flexible and adaptable within an overall integrated framework. Senior managers have a key role in creating this integration, by providing employees and others with a mental picture of the organisation, giving them a feeling of stability in the midst of flexibility and change.

The structure of the new organisation stands out in sharp contrast to that of the traditional Taylorist one, which is characterised by "Standardisation" or "Standardised Integration". The Taylorist organisation is one in which standardised goods were produced in large volume by standardised production line techniques for markets which were also seen to be largely standardised. The workforce had to be disciplined by an autocratic management to perform simple and endlessly repeated tasks in vast plants. The model followed in this type of organisation was a closed, mechanistic, inward-looking one which was not expected to change from week to week.

People merely had to learn to do their own job, and the more they carried this out in a mechanistic way the better. According to this scenario, learning was seen as something mechanistic and procedural, which people went through at the beginning of their working lives. Their minds were "set up" to carry out specifically determined activities. Occasionally they needed an overhaul and this was considered retraining. In this mechanistic model of learning the main emphasis was on psychomotor skills and the learning of rules and procedures, drawn up by the elitist designers of the mechanistic system. The learning act therefore consisted in taking on someone else's view of how you should do your job. This kind of learning environment can be classified as a "passive learning environment" dominated by a transmitting approach.

The paradigm of the "new" organisation relates to a system which is flexible and at the same time integrated - a "Flexible Integrated System" This means an organisation which is decentralised, made up of autonomous work groups, giving greater responsibility to individuals, while at the same time integrated by common values and agreed strategies. The latter creates a corporate culture which holds the organisation together. This organisation must be open "outwards" to the external environment in relation to markets, and political, social and financial issues as well as being open "inwardly" through a human resources policy which emphasises trust, responsibility and initiative. This organisation, therefore, has more "surface" exposed to the outside environment, is market-driven and is able to move very quickly. It operates in a growing international environment so its products have to be "world class" in terms of quality and cost. Internally the style of management is "horizontal". The company is "flatter" with fewer layers of management. The role of the Chief Executive in offering leadership and helping people to understand and agree to work toward common goals becomes more important. (Management for the Future,1988).

Ironically the new kind of organisation demands the kind of management which implements policy acceptable to the left and right wings of the political spectrum. A business policy which emphasises the market forces and competition is in line with rightist thinking, while a human resources policy promoting self-management, responsibility and team work is in line with leftist thinking. (See "How the Company Man Becomes a Leftist Hero" Sunday Times, London, 29th October,1989)

New Qualifications

The general workforce also needs a dimension of business/enterprise attitudes. The "new" qualifications, covering such areas as initiative, responsibility and working in a team, describe a person who is very different to the Taylorist one. These new qualifications or competencies have been called "Les Competences de 3eme Dimension"(Aubrun et Orofiamma, 1989). The new kind of person has also been referred to as the Self-Managing Person. A Self- Learning Competency is one of the key attributes of the self-managing person. The learning environment which promotes this kind of person must be a dynamic one which facilitates learning in and through work. Companies must create and plan this active learning environment. In that way they become self-learning or self-qualifying organisations.

3 CIM and the Need for Human Competence and Learning Ability.

Advances in new technology over recent years have provided new vehicles to further the attainment of the flexible integrated organisation, so the term "Computer-integrated Manufacturing" has emerged. How can an organisation and the individuals in it learn to use new computerised technology to develop an integrated and flexible company ? The answer to this question presupposes another question: What is the most effective way to use computers in creating integration and flexibility in the new manufacturing organisation ?

Ebel (1989), presents a state-of-the-art answer to the above questions. Based on an analysis of pilot projects and accompanying research in Japan, the United States and Europe, he describes the two main trends as being the "technocentric" one and the "human-centered" one. The technocentric approach which is to be found in its purest form in the United States, and which has been transferred to some other countries, constitutes "... an attempt gradually to reduce human intervention in the production process to a minimum, and to design systems flexible enough to react rapidly to changing market demand for high-quality products. Workers and technicians on the shop floor are typically seen as unpredictable, troublesome and unreliable elements capable of disturbing the production and information flow, which is best controlled centrally through computers. The unmanned factory is the ultimate goal. Only a residual role is assigned to workers, whose skills are supposed to be incorporated gradually and progressively into the machines." (Ebel, 1989, p.536-537)

In the technocentric approach integration is achieved primarily through reliance on the flexible expert designed technology, and not on what is seen to be the inflexible human factor in the company. This approach, whose earlier stages of development are described by Shaiken (1984) and which is a reformulation of the Taylorist scientific philosophy in the age of the computer, obviously gives a minor role to the concept of self-learning individuals, and indeed to any degree of comprehensive training and development in the workplace.

EUROTECNET findings and other insights outlined by Peters (1988) raise serious questions about this approach on the grounds of cost and efficiency. It is being recognised that because the flexible automation central to CIM is so complex in relation to hardware and software, it is liable to breakdown. Highly skilled and flexible (self-directing) workers are therefore required to supervise and control the overall process. The cultural consequences of planning a society which is under the control of, rather than controlling technology is a critical issue that has received much comment. (Postman,1988).

The Japanese take the human factor as their starting point in relation to CIM. In their gradual incremental ("Kaizen") approach to the introduction of advanced manufacturing processes, they see the most flexible element as being the highly skilled and motivated workforce. "Quality" in fact is often understood to refer to the quality of the competence of the workforce, that is the "quality of the process", rather than the quality of the end product or "result-oriented quality" (Imai, 1986).

In general in Europe, too, the human centered approach is the dominant one. Companies, particularly small and medium-sized enterprises, adopt a pragmatic gradualist approach to CIM, relying by and large on the skilled craftsman to provide a flexible and customised service to their clients. "The new computerised flexible and integrated automation equipment is primarily seen as an improved tool in the hands of a skilled and versatile workforce serving to enhance existing know-how and to permit greater flexibility, higher productivity, better product quality and shorter delivery times." (Ebel, 1989 p. 538). Cooley (1988), makes the point that we are now at an unique historical turning point in relation to how we design the workplace of the future.

The technocentric view is being questioned for one reason it has not produced results. Increasingly a more human-centered perspective on CIM is emerging as the dominant view and the simplified and instant solutions offered by some designers of "CIM packages" are seen to be invalid. Studies by Martin (1988), and Brodner (1987) bear this out. CIM is now seen to be a strategy centered on business and just not centered on technology. There is a need for a "people-at-work revolution" to match the "technological revolution". The technological advances made must be utilised in a manner to maximise overall efficiency by means of more flexibility, decentralisation and autonomous work groups. There is a need to integrate the latest technological and organisational approaches to achieve business objectives. Because CIM is a Business Strategy (inside peoples heads), linking different specialisations and departments, the need for an integrated organisational and competence development approach is also seen as crucial.

Organisations therefore need to set up efficient integrated and flexible management and organisation systems before they "computerise" them. Ebel notes that a number of surveys point out that " ...manufacturers who have introduced advanced manufacturing systems attribute between 40% and 70% of the total improvement achieved to organisational changes. In other words, the main benefit does not necessarily stem from sophisticated and integrated technology itself but from the reform of management practices and from a more transparent and efficient organisation" (Ebel p.542). Thus the technological strength made available by CIM can be taken advantage of by reinforcing the human factor rather that lessening it.

4   "New Knowledge": "New" Way of Learning.

Much discussion has taken place on the nature of new knowledge. This "Complex Systemic Knowledge" combines the theoretical and the practical, the general and the specific, the organisational and the individual, the human and the technological. A great debate has also taken place concerning the nature of the teaching and training systems to enable people to learn it. This is an issue for both initial training and continuing training. Interest is now being focused much more on the nature of learning from the "learners" point of view, and there is a revival of interest in "natural" on-the-job learning. In many ways this is a reinterpretation of the traditional "apprenticeship" model of learning. So perhaps the new way of learning, or self-learning, is not so new after all ! There is a lot of truth in what Postman (1988, p.18) says about the purpose of social research being to rediscover the truths of social life. He states that "...social research never discovers anything. It only rediscovers what people once were told and need to be told again".

The interest in the "new" way of learning has come about because many commentators in the business world, and in the fields of education and training, realise that the present systems are not enabling people to learn the "new" complex socio-technical knowledge, which has practical and tacit as well as theoretical sides to it. This problem has been noted by the European Round Table of Industrialists, (1989). There is also a criticism of the overdominant position held by the "objective expert scientific view of knowledge" as compared to "knowledge derived from practice" viewpoint, which has many subtle contextual nuances that do not fit into the observable and measurable categories demanded by objective science. (Fragniere, 1976,p.9; Schon, 1983 ; Cooley, 1988).

The modern professional worker is seen as someone who has the competency to deal effectively with situations in their entirety, "the competency to act". Self-Learning Competency is one of the "core qualifications" which is needed by this worker.

The new workers require mental skills rather than manual skills. This means that they must be able to form "mental images" when working with automated machines. They must be multi-skilled and be capable of explaining to their colleagues what they are doing. (This communication process is in fact a joint self-learning process). They must be capable of understanding the interrelationship between their actions and the overall success of the organisation.

According to Ebel (1989), the kind of knowledge required by people working with CIM is described as follows: "CIM requires...people who understand production methods and the system and are capable of handling a great deal of technical information and of taking decisions on the spot" (Ebel,p.545) "... the qualified, motivated and experienced worker familiar with the system, can cope with uncertainty and assess situations, find and interpret faults rapidly and correct them. Judgement backed up by technical knowledge and experience, understanding of the system and common sense is a human quality that cannot be replaced by computers or artificial intelligence in the foreseeable future. In CIM systems, machines and computers may well take over most routine and physical tasks but they do not relieve the people involved from thinking, critical decision-making and responsibility" (Ebel p.543).

The ability to operate within a human system, to manage one's role in a flexible way, as distinct from carrying out the static role that has been assigned, is required by the skilled CIM worker. The concept of "managing yourself in role" which led to the idea of "Organisational Role Analysis", is very relevant to this topic (Reed, 1976,1985).   This kind of knowledge can be described as that which allows a person to answer the following question in a satisfactory way: How do I fit into the group? The hallmark of the new organisation is human "integration" and interdependency. The overview is becoming much more important. The links between individuals and the overall system as well as between technology and people have to be carefully considered. All people in the organisation have to work towards taking an overall integrated view of things. This call for a new style of learning by everybody in the organisation in the context of the new organisational and management policy already described.

Active Learning Environment

People will not learn in an active self-learning manner unless they are operating in an active learning work environment. Self-learning does not mean that people learn solely by trial and error without instruction and guidance from instructors. The active learning environment is a social reality which has to be designed and planned. The instructors still have a key but a changed role. There is also a critical role for the line managers to facilitate learning as part and parcel of the everyday production process. The motivation for learning, which gives the learner the answer to the "why learn?" question, lies in addressing real problems, so the on-the-job situations must be utilised as prime learning opportunities. The line supervisor or manager becomes the key person to make sure that these learning opportunities are utilised.

The features of the active learning environment are as follows:

- The relationship between the learner and the instructor is a democratic one rather than a hierarchical one

- The instructor/line manager is a partner in learning rather than a domineering expert

- The instructor takes on the role of guide, counsellor and facilitator.

- Learning is "project-based" or "action-field" centered.

- Learning contracts are drawn up between the learner, the instructor and the company.

- People should be given maximum opportunities at work to seek information by themselves and to transfer what they have learnt to new situations.

- Team learning takes place with people at different skill levels learning together.

- Reflection meetings are organised

To encourage people to learn new skills they should be rewarded for what they know and not for what they do.

Use of computer systems based on embedded knowledge are used to assist learning rather than those which transmit "expert knowledge" or decisions in a passive way (Nyhan, Essen Report, 1989).

New terminology has emerged in different European countries in recent times to describe this new kind of learning environment. In the U.K. the term "learning company" has been used to define "an organisation which facilitates the learning of all of its members and continuously transforms itself" (Pedler, Boydell and Burgoyne, 1989). The term " Qualifying Organisations" has emerged in France (Riboud, 1985). In Germany the term "Learning Oriented Work" has appeared (Lellmann, 1989). This active learning environment is also referred to as "the focused approach to training and development", in the sense that it focuses on organisational and individual goals. The learning steps and methodology are planned to deal with organisational needs which everybody in the company is aware of. This is contrasted with a fragmented (chaotic learning by doing) and a formal (passive) learning environment. (See Management for the Future, 1988, and Nyhan, 1991)

This kind of learning environment cannot be introduced into a company overnight. It takes time and it needs to be planned strategically by top management so that it is integrated into the business goals and culture of the company. The new learning environment must be introduced in line with the flexible organisation concept, in which skilled workers are given more autonomy and middle management and supervisors take on new roles. The importance of the role of the Chief Executive in articulating the nature of this flexible learning environment which is integrated on the basis of well-defined and agreed organisational goals, cannot be overestimated. The Head of Training in one of Europe's largest companies commented that many organisations move between the two extremes of the fragmented (chaotic) learning system and the formal (passive) one, without ever establishing a strategically focused learning policy.

The introduction of an active learning environment needs to be urgently tackled however in the pragmatic and social interests of business growth and prosperity. This demands a long term learning-oriented strategy. It also calls for a belief in the growth potential, the dignity and creativity of people. By building up the Self-Learning Competency of everybody at work, the seeds of continuous learning and skill development will be sown. In that way the benefits of CIM and the future technologies still to be developed can be reaped to the full.

References

1. Argyris, C., Schon,D.A. (1978): Organisational Learning: A theory of Action Perspective, Addison Wesley, Reading, Massachusetts.

2. Aubrun, S., Orofiamma,R.: Conservatoire National Des Arts Et Metiers, Juillet, 1989

3. Brodner, P., (1987): Towards an anthropocentric approach in European manufacturing., CEDEFOP, Vocational Training Bulletin, No. 1, 1987.

4. Checkland, P., (1981): Systems Thinking, Systems Practice, Wiley, Chichester.

5. Cooley, M.: Creativity, Skill and Human-Centered Systems., in Goranzon., B., and Josefson, I., (Eds.), 1988, Knowledge, Skill and Artificial Intelligence., Springer-Verlag, London.

6. Ebel, Karl-H.: Manning the Unmanned Factory, International Labour Review, Vol.128, No.5, 1989.

7. European Round Table of Industrialists, (1989), Education and European Competence.

8. Fragniere, G., (1976): Education Without Frontiers, A Study of Education from the European Cultural Foundation's "Plan Europe 2000"., Duckworth, London.

9. Goranzon, B., and Josefson, I., (1988): Knowledge, Skill and Artificial Intelligence., Springer-Verlag, London.

10. Gullers, P., (1988): Automation, Skill, Apprenticeship, in Goranzon, B., and Josefson, I., Knowledge, Skill and Artificial Intelligence, Springer-Verlag, London.

11. Heidack, Clemens. (1989): Lernen der Zukunft, Lexika, Munchen.

12.   Henderson, Jeffrey,(1989): The Globalisation of High Technology Production, Routledge, London.

13.   Imai, Masaaki, (1986): Kaizen, The Key to Japan's Competitive Success., Random House Business Division, New York.

14.   Lellmann, D., (1989): Learning Oriented Work On The Job., EUROTECNET Newsletter, No.10, July 1989.

15.   Martin, T. (1988): The Need for Human Skills in Production- The Case of CHIM, Kernforschungszentrum Karlsruhe; (Paper presented at Venice R.S.O. Conference, October, 1988.

16.   Nyhan, B., (1989): Self-Learning Competency- the Key to life-long learning, EUROTECNET.

17.   Nyhan, B., (1989), Essen Seminar on Self-Learning Competenc: Report, EUROTECNET,

18.   Nyhan, B. (1991): Developing People's Ability To Learn, EUROTECNET/ European Interuniversity Press, Brussels.

19.   Pedler, M., Boydell, T., and Burgoyne,J.: Towards A Learning Company, MEAD, Journal of the Association for Management Education and Development, Vol.20, Part 1, 1989.

20.   Peters, Tom, (1988): Thriving on Chaos, Handbook for Management Revolution., Alfred A. Knopf, New York.

21.   Postman, Neil,(1988): Conscientious Objections, Stirring Up Trouble About Language, Technology, and Education., Alfred A Knopf.Inc., New York.

22.   Reed, B.D., (1976): Organisational Role Analysis, in Developing Social Skills in managers, Cooper, C.L. (Ed.), Macmillan, London.

23.   Reed, B.D., (1984), Stress: The Individual and the System., Educational Management and Administration 13.

24.   Riboud, A., (1985): Modernisation, mode d'emploi., B.S.N. 10/18

25.   Shaiken, Harlry, (1984), Work Transformed, Automation and Labour in thr Computer Age, Holt, Rinehart and Winston, New York.

26.   Schon, D., (1983), The Reflective Practioner, How Professionals Think in Action., Basic Books Inc., New York.

New Methods of Interdisciplinary Education for Engineers and Computer Scientists in the Field of CIM

Prof. Dr. Klaus-Jürgen Peschges, Jürgen Bollwahn, Kai Maurer,
Erich Reindel and Jörg Schumacher
CIM Project at the Fachhochschule for Technik, Mannheim
Mannheim Germany

Abstract

To improve interdisciplinary education, a CIM project of the Land Baden-Württemberg was started in 1988. The project (directed by the Fachhochschule für Technik Mannheim) was funded by the German Federal Ministry of Science, the Baden-Württemberg Ministry of Science and Arts and eight industrial companies mainly from the Rhine-Neckar area. As a result a computer based interactive multimedia teaching and learning system (abbreviated LLS) has been developed to impart CIM-related knowledge. Target groups of the LLS are students of engineering and computer sciences and employees on the job concerned with CIM matters.

LLS is not a pure self learning program but serves as the theoretical section of CIM education, which is taught through seminars under tutorial guidance. The practical part of CIM education is undertaken at the FHT Mannheim through the PROCIM project, in which an electro-mechanic environmental data scanner is being developed by all faculties in interdisciplinary cooperation.

1   Introduction

With the current trend of university education focusing on the training of specialists, the importance of cooperative thinking and interdisciplinary methods is often overlooked; this lack of lateralism is brought into professional life and industry later. But the ability to recognize and understand problems from other's points of view is a necessary requirement for solving common difficulties. Apart from the basic qualifications needed by specialist engineers, knowledge of other disciplines should be taught to increase flexibility within engineering professions.

Pure specialists are overtaxed by presently developed highly integrated and complex methods of working, which can be summarised under the "CIM" label. A big problem in this connection is the realisation of CIM in different divisions of a company in rejection of Taylorism.

Besides expert knowledge, general capabilities are important for any staff member in any job profile. These skills are necessary to help the employee to understand complex processes and make decisions on a basis of a general understanding of all relevant facts.

Because of this challenge, Baden-Württemberg supports a CIM project headed by the Fachhochschule für Technik Mannheim. The paramount object of this program is to find new ways towards interdisciplinarity and to make it accessible for both students and engineers on the job. The only way to achieve success is if those who develop new curricula and teach CIM proceed in a cooperative manner with "CIMed" brains.

Besides the Fachhochschule für Technik Mannheim (project management) other universities, polytechnics, vocational academies and leading companies of the Rhine-Neckar region are involved, especially in the development of the LLS (which is the abbreviation for the commonly created teaching and learning system). In a unique common approach this interactive training system in the field of CIM has been realised for the two target groups - industry and higher education. The use of LLS for further education of students and employees is planned for the future.

Simultaneously the CIM net has already been thrown over the FHT Mannheim through the transfer of CIM structures into the daily work of teaching staff and students. With the use of LLS students can be acquainted deliberately to CIM during theoretical education. CIM will be a firm element in the curricula of all FHT faculties. Thinking in complex structures will also be taught by the LLS, and after an introduction to the CIM components the application of the learning matter on concrete examples will be possible.

To avoid losing reference to reality and apply the contents of the LLS, a new product, the Modular Environment Measuring Station (abbreviated MUMS), an electro-mechanic scanner of environmental data, has been developed and is manufactured and improved by all faculties of the FHT. An indisputable advantage of MUMS is that theory is supported by practical exercises leading to greater learning success. The development of MUMS is the practical component of CIM education at the FHT Mannheim.

In the same way that the development of MUMS demonstrates an efficient collaboration of the separate faculties, the realisation of the LLS is proof of unique cooperation of industry, on the one hand, and universities and polytechnics, on the other. Even the Fachhochschule für Gestaltung (FHG), an college for art and design, is involved in MUMS development: another good example of interdisciplinarity.

Thanks to all collaborators (professors, assistants and students) of the Fachhochschule für Technik and the Fachhochschule für Gestaltung.
Special gratitude to the initiators of the CIM-LLS project:

Prof. von Hoyningen-Huene, rector of the FHT Mannheim,
Mr. Köngeter, Mercedes-Benz AG Mannheim,
Prof. Dr. Sommer, Pädagogische Hochschule Karlsruhe.
Also gratitude to:

"Forschungsvorhaben CIM und computer-unterstützte interaktive Medien" (CIM research programme and computer-aided interactive media):
Mrs. Ingela Jöns, Mrs. Barbara Kirchner and Dr. Sandor Vajna
and to
Ministerium für Wissenschaft und Kunst (Ministry of Science and Art) of the State of Baden-Württemberg, Stuttgart,
Bundesministerium für Bildung und Wissenschaft (Federal Ministry of Education and Science), Bonn.

Involved companies:

ABB AG Mannheim, AEG AG Frankfurt, BASF AG Ludwigshafen, Bopp & Reuter GmbH Mannheim, Deere & Co. Mannheim, Carl Freudenberg Weinheim, Mercedes-Benz AG Mannheim and Pepperl & Fuchs GmbH Mannheim,
and to
Mr. Klaus Bessau (rector of the FHG Mannheim).

2 The Target Group of the CIM-LLS Training Program

At the very beginning of the project the target groups of the CIM education program had been defined as follows:

-       Managers, engineers and employees with similar knowledge, whose concern is to plan and realise CIM projects.
-       Students of mechanical, electrical and economical engineering and computer scientists.

The demand for CIM training covers hierarchical (vertical) levels as well as different scopes of duty (horizontal levels).

Following the integration concept of CIM, the education program should not aim at any singular (technical) target group. Moreover, persons from different faculties should be integrated in the target-groups and the CIM promotion teams should consist of persons from a variety of horizontal and vertical levels. Students of different sciences are included, too.

As a result of exhaustive discussion about how to define the target group, the following statement has been formulated as,
"The groups of staff members who, in fact, have influence on other departments of a company now and in the future. Prospective staff members include students of the specific faculties."

According to this extended definition of the target group, the teaching and learning system LLS has been developed so that it can reach maximal flexibility. As a result, every company is able to select learners for its own purposes and students of different faculties may use the LLS as well.

The heterogeneity of the target-group had to be considered when developing the

LLS because for effective education it is necessary to build up on pre-existing knowledge and to incorporate mental structures of the learner. Also, the motivation to learn has to be kept alive during the whole learning process. Therefore, while the content of the learning matter has to be held at a medium level of newness and complexity, it should be challenging, too.

3   Content of Learning Matter - The CIM Problem

For a long time CIM was considered from an exclusively technocentric viewpoint with the aim of setting up an unmanned factory. The CIM idea as well as CIM training efforts were mainly reduced to questions referring to the interconnection of software and hardware systems which were mostly applied in traditional Taylorist work environments. These environments are characterised by a strictly department-oriented way of working and thinking with a low degree of interaction between departments or corporate functions, a low degree of flexibility, long-term deliveries, high production costs and intense efforts in quality control. The introduction of CIM components into this kind of work organisation failed in most cases and was financially unacceptable for innovative industrial enterprises.

From today's viewpoint CIM describes the computer-supported integration of all corporate domains or functions, from market analysis to recycling of a planned product. And it must be underscored that the new approach instead of concentrating on hardware and software issues places strong emphasis on the concept of integration which should manifest itself in the staff's behaviour and thinking before interconnecting computer systems and databases.
This future integrated way of producing is expected to deliver products

- of higher quality,
- at lower production costs,
- in shorter terms of delivery and
- having a higher diversity of product variants with fewer numbers of pieces.

As a consequence the workforce will need an essentially higher degree of training which should correspond to the high complexity of the interconnected work processes and which takes into consideration the whole spectrum of activities accompanying the production process, from market analysis to the recycling of the product, instead of a restricted department or task oriented view of activities.

The following diagram gives an idea of the quality of knowledge/training of this future CIM workforce compared to traditional staff:

-       Staff in traditional work environments are characterised by a high degree of expert knowledge related to their tasks/functions but a low degree of general knowledge concerning other tasks or corporate functions.
-       In integrated CIM work environments the workforce (CIM staff) needs, in addition to a high degree of expert knowledge, a high degree of general knowledge which should enable staff to be conscious of the direct impact of their decisions on other corporate functions.

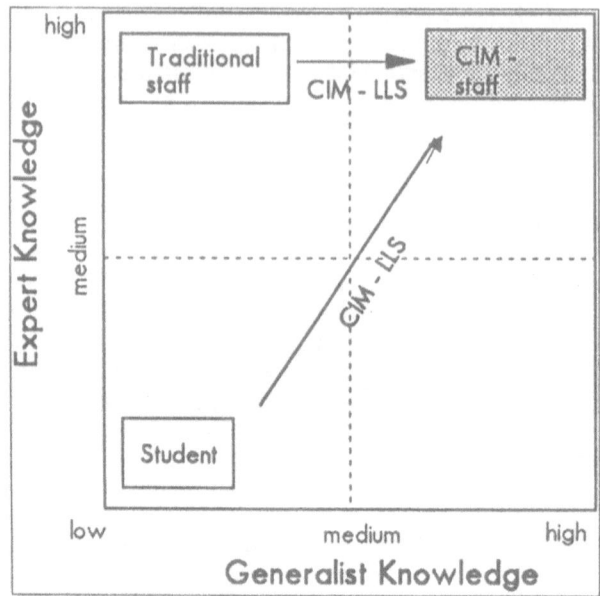

Fig. 1   The idea of the quality of knowledge/training og this furture CIM

The success of CIM will ultimately depend on early and adequate training for these new tasks.

In order to implement the "CIM philosophy" successfully in the industrial world there is a need for a fast global introduction of the CIM concept into the companies at all functional and hierarchical levels and into higher technical education where the future workforce is educated.
This leads to a gap, however:
The required amount of instructors in industry or of professors in higher technical education (universities and technical colleges) are neither available, nor does the expertise in this domain of knowledge already exist because it is too recent.

A computer based, multimedia training system was developed to bridge this gap. It provides a unique and authentic base of knowledge developed by a group of experts from industry and higher education.

4   Didactic Information about the Course Structure

4.1   A Neutral CIM Model

The main functions and duties of other company departments have to be known to fulfill a task. Moreover, one has to recognize all dependencies and relations which could even indirectly have an influence on one's function. Therefore, a sector- and company-independent CIM model has been developed which allows representation of each kind of company - no matter how big it is or what they are it produces -

and also serves as the structural content of the learning system. So a common educational concept applicable to different companies has been worked up and in the following picture this general structure of CIM training is shown. Each kind of company can be defined by the following functions:

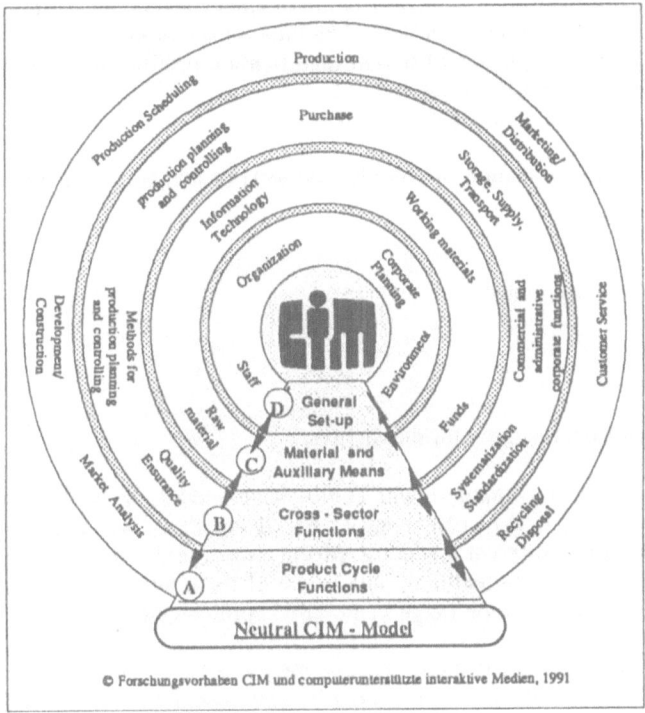

Fig. 2   The different transferable education concept

- Product cycle functions which are passed through during a product cycle.

- Cross-section functions which accompany a product cycle.

- Material and auxiliary means for planning and realisation of a product cycle.

- Company and non-company general set-up such as staff or environment having direct impact on all the functions mentioned.

In this model the complex dependencies and relationships within a company can be shown in a structured manner and the learning matter is now more easily transferable to a computer-based learning system.

4.2 Content of the Teaching and Learning System LLS

The complexity of the subjects to be learned as well as the heterogeneity of the target groups, composed of students from different academic faculties and the workforce from different branches or hierarchical levels and corporate functions, require a flexible training tool that takes into consideration
- the different prerequisites (different qualification profiles and degrees) as well as
- the individual learning styles and
- learning progress of the learners and

supports the dissemination of complex subjects or contents supported by

- graphics,
- animation,
- still and moving image,
- sound,
- text and
- simulations

in permanent interaction with the learner.

The linking up of computers and audio-visual media, as in the interactive CIM Learning System (CIM-LLS), meets these demands. CIM-LLS combines all the advantages of the different media (or information types) and enables optimisation of the learning process. In addition, on account of these different navigation paths, the system offers quite a wide range of possible applications.

The application of the learning system ranges from a "pure" information system with a relatively low level of learner system interaction to a highly interactive self-learning medium.

The consortium favoured the idea of dividing the learning process into two phases in which the learning system is integrated. Meetings of all learners supported by tutors and self-learning at the computer. This should be applied in the curricula and training events of both industry and higher education.

Almost all of the identified corporate functions are covered by learning modules. So-called "Interface Modules" explicitly describe the dependencies and interrelation of different corporate functions. The next Figure (Overview: CIM Learning System) gives an overview of the learning modules and it illustrates the structure of CIM-LLS and the CIM model.

In sum, the whole system consists of more than 100 learning hours and includes

- a film on videodisc of a total length of 30 minutes which can be applied as a motivating introduction to the CIM problem,
- two introductory modules explaining the handling and structure of the learning system and
- twenty-three learning modules which can be accessed freely.

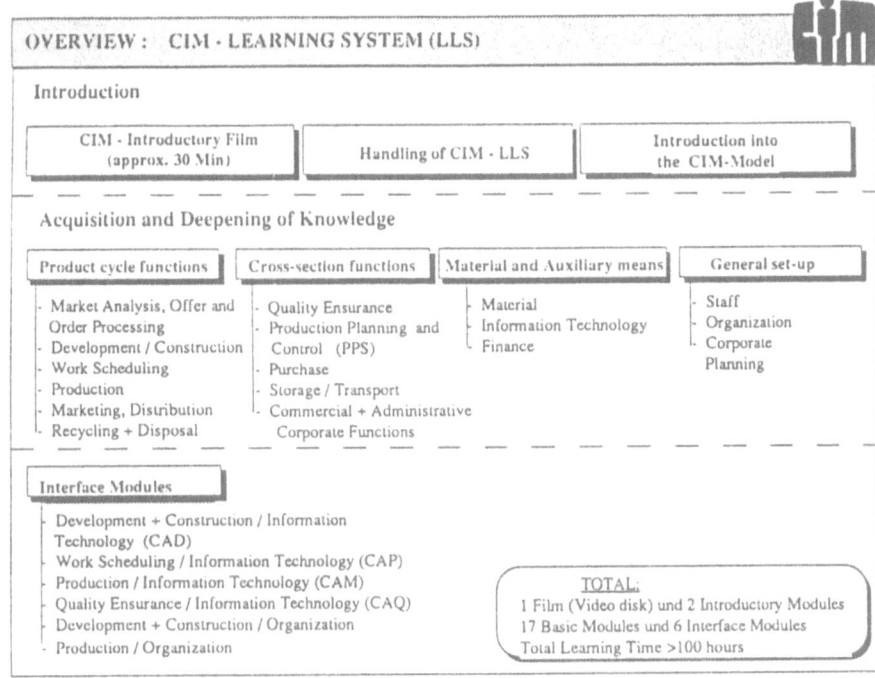

Fig. 3   The CIM learning system

5   Methods and Organisation of Teaching and Learning

At the Fachhochschule für Technik Mannheim (FHT) a committee has been established which coordinates CIM. It consists of one professor from each faculty, the head of the computer centre, the rector of the FHT and the chairman, Professor Peschges. The goal is to work out exemplary CIM solutions in teaching and research at the FHT. Therefore, a general concept is used which integrates all faculties and generates a "CIMed" FHT. It is intended to involve all departments of the FHT and the specific know-how in the created CIM structure. Integration problems should be reduced by means of interdisciplinary collaboration. But it has to be taken into account that "insular" solutions will not be created. Rather, "faculty wide" integration is aimed at: This is much more important than a CIM net within a single faculty of the FHT.

As a result of the CIM project, those organisationally and spatially separated faculties will be given the possibility to coordinate their activities via a network of information and by use of common databases.

The funding of the purchase of the software and hardware components is determined solely in accordance with the requirements and capacity of the user. It is advantageous that expense decisions are made on the basis of user requirements as they arise rather than via a pre-determined purchasing plan. If possible, a level of standardisation or compatibility of technology equipment is sought, though specific user needs are not sacrificed for "homogeneity". This would lead to a "CIM model enterprise" which would both be an artificial working environment and would hide specific problems of CIM.

The LLS (teaching and learning system) is the main tool to impart the basic knowledge of CIM. This means self-learning, supported by multimedia teaching software, as well as feedback from this interactive teaching and learning system through seminar-like events for student groups, which consist of persons of all faculties.

Simultaneously the FHT staff and the students are able to transfer theoretical CIM knowledge to practical work in development and improvement of MUMS. All persons involved, staff as well as students, qualify themselves by this mixed theoretical-practical approach for CIM. For engineers in industry the practical part of CIM education can be found in their companies but for students no such possibility existed. Something had to be done in this field.

5.1 The Interdisciplinary Practical Part of CIM Education at the FHT

It is characteristic of universities and technical colleges that they support singular activities of the institutes and faculties with no connection between each other. To overcome such a separation of knowledge a common aim is necessary which can be reached only via joint projects and products. This is similar to companies in which the different departments and divisions are forced to work together. The following passages describe the different points of view on this concept. Students are predominantly concerned with CIM during their examination and thesis work, while professors and assistants are more involved in development, research and teaching in the field of CIM.

5.1.1 A CIM Product: The Modular Environment Measuring Station

On the basis of example, students are to understand better what CIM means, but at the very beginning in 1988, when the CIM concept had been worked out, it seemed to be too early for a commitment. It was believed that CIM implementation could be reached by an abstract approach at a technical level. But since difficulties arose, it was decided to develop a product which allowed the CIM integration of all faculties more easily and to prevent divergent activities.

A product which is appropriate for CIM demonstration within the FHT has to comply with certain conditions, e.g. it has to be a electronic-mechanical product, so that all faculties of the FHT can be involved in the development of such a product. Moreover, it has to be sellable and suitable for adjustment to a customer's specific requirements and it must involve some parts and material provided by outside companies. These preconditions enable the FHT to act like a real company with all the characteristics concerning CIM.

The Modular Environment Measuring Station (MUMS) was designed to be consistent with the above conditions. MUMS is a scanner for environmental data, e.g. air humidity, temperature and pressure. A working team, consisting of persons from different faculties, was assigned to come up with a concept. Even the Fachhochschule für Gestaltung, an academy for art and design, has been involved. By means of "methodological design" and "simultaneous engineering", a basis for the final MUMS was attained.

Following a product cycle, different parts of MUMS (e.g. injection moulding of the cabinet or circuit board) were completed by single groups (consisting of professors, assistants and students). Essentially MUMS consists of four components: cabinet, circuit boards, sensors and software for data processing.

Those four basic elements run parallel but with steady feedback throughout development and manufacturing and are finally assembled into a unit. But that is not the end of the MUMS cycle at the FHT Mannheim. At present some teams are working on the recycling of the plastic cabinet or old circuit boards.

In the first stage of the development, it has been demonstrated successfully that air temperature, pressure and humidity, basic data of every environmental measuring technique, can be scanned by MUMS. So far funds have been granted to build 20 MUMS containing the basic modules mentioned above.

Further development of MUMS has been planned. The next stage involves the measurement of air pollutants, e.g. nitrogen oxide and ozone. This should be done in the way defined by CIM so that all the specialist potential can be utilised. The aim is to integrate MUMS into an early warning system for air pollution. This means a high-tech mobile laboratory capable of precise scanning and measurement of environmental data, supported by a network of MUMS spread around the scanned area. These will relay pollutant threshold alert values back to an information centre from which the mobile laboratory is sent to critical places.

5.1.2 The Organisation - The PROCIM Committee and Its Activities

The FHT Mannheim is split up into eight faculties and eleven courses of study. The teaching staff consists of about 90 professors, supported by their assistants, and part-time lecturers from industry. Besides looking after students' needs, the assistants are assigned to maintenance of all equipment units concerned with CIM and bought with CIM funds.

Because students should be involved in CIM in a realistic manner, a great deal of development and realisation projects are to be carried out by themselves mainly through papers and thesis work.

Since the collaboration of all faculties (professors, assistants and students) is needed - as defined by the CIM idea - a central instrument for coordination has to exist. Therefore, the PROCIM committee (in the following abbreviated as PROCOM) is in the centre of all activities concerning CIM at the FHT Mannheim. Professors, assistants and students of different faculties and of the Fachhochschule für Gestaltung are members of this working team. They coordinate CIM activities and promote CIM, the improvement of MUMS and the introduction of the interactive learning system. They are also responsible for presentations, e.g. during the annual information day of the FHT. Once a week the members of PROCOM come together to inform and to be informed about the present state of the CIM project. Also PROCOM decides what to do next. So feedback is possible for everyone about his activities concerning CIM and flexible action can be made. This is what is called "Simultaneous Engineering".

The "information policy" is one of the duties of the PROCOM chairman and his assistant. Every relevant fact is taken down in the minutes of the weekly PROCOM sessions. Those minutes are the main element of a chronological documentation of CIM proceedings. Cooperators are given the chance to be informed about "what's going on". Moreover, members who were absent during a PROCOM session can get information about the project efficiently.

Since the minutes are not as detailed as may be advantageous for some purposes (only rough facts are written down) PROCOM publishes a paper called MUMS-News, in which some problems (and their solutions) are described with more precise background knowledge.

In the entrance hall of the main building is a display case placed at a central point where up-to-date information about PROCOM, MUMS and LLS can be found. A CIM station has been established in the library where magazines, papers and books containing CIM information are deposited.

The heart of CIM information is the CIM centre. Everything concerning CIM, MUMS and LLS is collected there and all information is accessible to interested persons at all times.

Finally the main functions of PROCOM within CIM realisation at the FHT Mannheim are outlined below:

1) Coordination of all CIM activities concerning MUMS and LLS.
2) The members of PROCOM are always well-informed about the state of CIM and try to spread the CIM idea all over the FHT.
3) Since PROCOM utilises skills of professors, assistants and students of different faculties it is the motivation behind integration and promotion of interdisciplinarity.
4) PROCOM coordinates and manages CIM development at the FHT.

5.2   The Interdisciplinary Theoretical Part of CIM Education

Theoretical and practical approaches of CIM depend on each other. Because of the complexity of CIM the necessary knowledge can be imparted only when the different parts of a company are known and their interdependencies are brought into the minds of the learners. This is the aim of basic CIM training, where the main information is made available to all learners in the same way by the self-developed multimedia computer system LLS. Although this large teaching and learning system exists, self-learning is only one part of CIM training. The next passage deals with the involvement of the correct use of the LLS in the entire curriculum.

5.2.1 The Project "CIM Teaching/Learning System" - A Historical Survey

The "CIM Lehr-/Lernsystem (LLS)" (teaching and learning system) is the result of the CIM project of Baden-Württemberg. The project which was directed by the Fachhochschule für Technik Mannheim (Institute of Technology Mannheim) was funded by the German Federal Ministry of Science, the Baden-Württemberg Ministry of Science and Arts and eight industrial companies coming mainly from the Rhine-Neckar Area. The total funding amounted to DM 3.5 million.

The CIM project, which was started in 1988 and successfully completed in 1991, demonstrates the successful collaboration of higher education and industry in the domain of research and development (R&D) of new subjects (CIM) and training methods (multimedia computer-based training system).

In a unique common and interdisciplinary approach an interactive training system for both domains - industry and higher education - was realised in the field of CIM.

The main objectives of the project were

- theoretical analysis of all relevant matters concerning CIM from the viewpoint of the participating institutions and organisations;
- definition of a neutral, sector- and company-independent CIM model as the structural basis of the interactive CIM teaching and learning system;
- development and testing of an interdisciplinary and interactive media system for CIM training in higher education and corporate training;
- development of original curricula for CIM training in industry and higher education.

5.2.2 Theoretical CIM Education

Efficient CIM education is a precondition to the introduction of CIM into a company successfully. The tool for theoretical education is the LLS. New knowledge can be imparted by the computer system and previously learned information can be refreshed.

The LLS should not be used as a pure self-learning medium: dialogue and discussion with other learners and the tutor are also necessary. In this phase of the learning process, after being introduced to one part of the learning matter, e.g. by the "quality assurance" module, the learners can discuss problems, ask the tutor and can thus obtain the ability to apply the knowledge to actual problems.

6 The LLS Configuration

The learning system is available for Apple Macintosh and IBM-compatible PCs.

On the basis of software criteria such as flexibility, modularity and different ways of knowledge linking and knowledge presentation (e.g. text, graphics, animation, sound, video), it was decided to use HyperCard on a Macintosh to develop the system. In addition, MacroMind Director was used to include complex and coloured animation in HyperCard, and CourseBuilder was used to realise some modules in colour. To include video of high quality and with a fast access time, we chose a laser disc, which was already familiar technology at the time we made our decision.

Although the computer world is dominated by IBM DOS and its compatible systems, the Mac and its multimedia facilities were the first choice for development because of the above reasons. Introduction of the LLS in industry required a DOS-compatible version, too. With the appearance of Toolbook (by Asymetrix) in 1991 a programming package was available which is a complete analogy of HyperCard, enabling automated transfer of 95% of the HyperCard program parts from Macintosh to PC. Therefore, the PC disk operating system DOS only has to be extended by including Windows.

System	Software
Macintosh	
Mac II si, 4 MB RAM (under Finder)	**System Software 6.07**
160 MB HD	**HyperCard**
13″HighRes Screen	(both delivered with Mac)
	CourseBuilder
	(for development)
IBM and compatibles	
PC 368, 4 MB RAM, 200 MB HD 14‴VGA	**DOS 3.3**
Screen (680*480)	**Windows 3.1**
Audio Card (SoundBlaster)	**ToolBook 1.5**
Auxiliaries for both systems	
Video Overlay (Screen Machine)	
Programmable Laser Disc Player (Sony LDP 1600 P)	
MacroMind Director (for development)	
MacroMind Player (for use)	

Fig. 4   Survey of the minimum configuration

To handle the enormous amount of data concerning the learning system, both Macintosh and PC need hard disk memory of at least 160 MB and an overlay card to include the video picture on computer screen. For both systems we use the ScreenMachine (by Fast Electronic) because of its programming facilities and its low costs. In addition to this, a special sound card (here SoundBlaster) is needed for the PC. The cost of each system (minimum configuration) is about 10,000 DM to 15,000 DM. The figure 4 gives a survey of the minimal configuration in both of the computer worlds.

7 Experience and Prospects

The concept and the realisation of interdisciplinary and integrational CIM education at the FHT Mannheim stand in contradiction to the hitherto existing CIM development from the technical viewpoint. But this does not meet the demands of the modern manufacturing process. The CIM project of the FHT Mannheim will bridge this gap. The amount of positive resonance proves that this way is the right one. Particularly activities concerning MUMS have already been introduced to a number of professors, assistants and students in conjunction with CIM successfully. Nevertheless, it cannot be denied that problems are left which have to be overcome in future because there are still "specialists" left who are not interested in any other subjects. But experience shows that even they will be convinced.

The planned simultaneous introduction of the LLS to six polytechnics in Baden-Württemberg (coordinated by the FHT Mannheim) should be pointed out. Besides its activities concerning higher education, the developer group intends to introduce the LLS to certain user associations, such as technology transfer centres. Thus a lot of information will be gained which is very useful for the improvement of the LLS and the development of papers and texts for use by tutors as well as by learners during the learning process.

Many national and international institutions in the fields of higher education, trade and industry have devoted attention to the above mentioned activities and have all expressed interest in using the LLS.

> "CIM - only those who recognize everything will recognize the entire wholeness"

Guided by this motto the FHT Mannheim is leading in

- the development of a new interdisciplinary and far-reaching CIM education program (practical experience using PROCIM and theory via seminars) and in

  the development and introduction of new media (LLS) and educational methods in higher education.

In this way a renewal in higher education is being achieved.

8 The Impact of CIM Training on Organisational Development

CIM strategies aim to maximise the competitiveness and flexibility of companies to meet new demands. CIM can be incorporated through two main strategies: the technically centered approach or the integrational approach. All the possibilities for "CIMing" a company are just a combination of both these approaches with specific variations. Though the organisational philosophies found in nearly all companies are technically centered, the integrational approach is the ideal strategy to introduce CIM into a company.

Obviously different strategies in the implementation of CIM lead to different impacts on management and employees of a company. However, the basic idea of integration can be described as follows:

- Man, technology and organisation have to be matched fitted to each other and technology is only a working aid.
- Rather than seeing a task only as a question of technology, view it from all angles.
- The human factor has to be considered as an integral part of any plan.

The technology has to support task fulfillment and has to be subordinate to man. In contrast to this, from a technocentric viewpoint man has to assist technology.

The organisational principles by means of which integration is achieved include:

- The organisation intended has to be defined first before technology is considered.
- The organisation has to meet economic demands and social demands through
  - possibilities for communication and cooperation,
  - possibilities for learning while working,
  - maintenance and improvement of training and
  - a minimum of autonomy.
- The provision of working and learning structures that give the staff stimulating and interesting duties and working conditions.

The main characteristics of such an integration-oriented organisation which prefers integration are:

- As little division of labour as possible, which means the creation of all-embracing work processes that provide for transfer of responsibility and competence.
- Cooperation of different departments.
- No centralised planning and controlling.
- Few hierarchical levels.
- Teamwork.
- High flexibility with a low level of standardisation and automation.

The preceding was only a short summary of the entire impact which CIM training has on organisation. A great deal more can be found in the LLS "Organisation" module. Finally, it has to be pointed out that CIM will change organisation very much indeed.

Appendix

Jürgen Broschat / Werner Kötter / Tilmann Krogoll
Zentrum für Wissenschaftlichen Gerätebau (ZWG)
Rudower Chaussee 6
O-1199 Berlin

Dr. Martin Burgmer
University of Dortmund
Department of Mechanical and Industrial Engineering
Chair of Technology and Didactics
Prof. Dr.-Ing. Udo Schüler
Emil-Figge-Str. 50
4600 Dortmund

Prof. Peter Fröhlich / Prof. Dr. Hans -J. Holland
Fachhochschule Wiesbaden
Fachbereich Maschinenbau
Am Brückenweg 26
6090 Rüsselsheim

Dr.-Ing. Martina Klocke
ÜAZ-Elmshorn
Ramskamp 8
2200 Elmshorn

Prof. Dr. Johan Vesterager
Associate Professor, PhD
Institute of Production Management and Industrial Engineering
Technical University of Denmark
Building 423
DK-2800 Lyngby / Denmark

Jörg Kluger/ Norbert Meyer/ Helmut Richter
Berufsförderungszentrum Essen e.V.
PTQ Project Group
Altenessener Str. 80-84
4300 Essen 12

Heiner Mählck
Panskus-Unternehmensberatung
BDU
Bismarckstraße 11
5600 Wuppertal 1

Dipl.-Kfm. Markus Nüttgens
Scientific assistant
Institut für Wirtschaftsinformatik (IWi)
Universität des Saarlandes
Director Prof. Dr. A.-W. Scheer
Altkesseler Str. 17
6600 Saarbrücken 5

Prof. Dr.-Ing. Walter E. Theuerkauf / Andreas Weiner
Institut für Angewandte Elektronik und Technikpädagogik (IAET)
University Hannover
Im Moore 11 A
3000 Hannover 1

Prof. Dr.-Ing. Wolf Martin
Institut für Gewerblich - Technische Wissenschaften
Universität Hamburg
FB 06/Institut 11
Von-Melle-Park 8
2000 Hamburg 13

Dipl.-Ing. Flavio Canetti / Dipl.-Ing. Reinhold Dressler
Siemens Anlagentechnik
c/o Siemens AG
Abt. ANL A 245
Postfach 32 40
8520 Erlangen

Dipl.-Ing Frank Hanewinkel
CIM-Fabrik Hannover GmbH (CFH)
Hollerithallee 6
3000 Hannover 21

Dr. Willi Petersen / Manfred Schön
Institut Technik & Bildung (ITB)
University of Bremen
Grazer Str. 2
2800 Bremen 33

Dr. Herbert Tilch
Institut Technik & Bildung (ITB)
University of Bremen
Grazer Str. 2
2800 Bremen 33

Dr. Kuteiba Alshahid / Roland Thomas
ERT (UK) LTD
95 Ditchling Road
Brighton, BN1 4SB England

Dr. Henk BOLK
InterVisie Consultants in Management and Strategy
Schipholweg 88
2316 XD Leiden Netherland

Prof. Dr. Friedrich Wilhelm Bruns / Lüder Busekros / Axel Heimbucher
Forschungsgruppe artec
University of Bremen
Bibliotheksstraße
2800 Bremen 33

Barry Nyhan
Senior Consultant
c/o EUROTECNET
Technical Assistance Office
Rue des Deux Eglises 37
B-1040 Bruxelles Belgium

Prof. Dr. Klaus-Jürgen Peschges / Jürgen Bollwahn / Kai Maurer / Erich Reindel /
Jörg Schumacher
c/o "CIM Lehr-Lernsystem der
FHT-Mannheim"
Speyer Str. 4
6800 Mannheim 1

Author Index